Distribution of Statistical Observables for Anomalous and Nonergodic Diffusions

Distribution of Statistical Observables for Anomalous and Nonergodic Diffusions

From Statistics to Mathematics

Weihua Deng

Xudong Wang

Daxin Nie

Xing Liu

CRC Press
Taylor & Francis Group
Boca Raton London New York

CRC Press is an imprint of the
Taylor & Francis Group, an **informa** business

First edition published 2022
by CRC Press
6000 Broken Sound Parkway NW, Suite 300, Boca Raton, FL 33487-2742

and by CRC Press
4 Park Square, Milton Park, Abingdon, Oxon, OX14 4RN

CRC Press is an imprint of Taylor & Francis Group, LLC

ISBN: 978-1-032-24521-8 (hbk)
ISBN: 978-1-032-24523-2 (pbk)
ISBN: 978-1-003-27909-9 (ebk)

DOI: 10.1201/9781003279099

Typeset in Latin Modern
by KnowledgeWorks Global Ltd.

Contents

CHAPTER 2 ■ Numerical Methods for the Governing Equations of PDF of Statistical Observables

Preface

The natural world is very complicated. Mathematics show unreasonable effectiveness in the natural world, most of the time based on physical and/or chemical principles. From microscopic field, e.g., macromolecules, including proteins, lipids, carbohydrates, nucleic acids, etc., which compose of organisms, to macroscopic level, e.g., pandemic of COVID-2019, the cause, the function, and the consequence are generally resulted by the motion/assembly of particles. Abstractly, the dynamical behaviors of particles can be modelled by stochastic processes, which more often are non-Brownian, because of the complexity of the environment. This is the topic the book focuses on. We integrally discuss the scientific problems ranging from stochastic and deterministic modelling, physical applications, to mathematical issues, including analyses of (stochastic) partial differential equations (PDEs), scientific computations, stochastic analyses, etc.

In fact, almost all the scientific problems cross several disciplines. It is not an exception for the anomalous and nonergodic diffusion. With the rapid development in the past decades, there are already some good monographs related to this topic, but as the current research paradigm, every research group has particular research background, resulting in every monograph treating a special subject. The PDEs governing the probability density function of the statistic observables, generally involve the nonlocal operators, which are fractional operators in the special cases. Because of this, a lot of people turn to the research on numerical method and well-posedness and regularity of fractional PDEs (mainly on fractional diffusion equations), so there are around ten monographs on numerical method for fractional PDEs and one or two monographs for the analyses of fractional PDEs. As a mathematic research topic, Lévy processes have been well studied. It should be a particular branch of the stochastic processes we consider. There are more than ten books on the Lévy processes, targeting the readers with strong mathematic background on real analyses and stochastic analyses. There are also two or three monographs writing from the perspective of statistics itself or statistical physics, which are not very suitable for mathematicians and computational scientists. Essentially, all these scientific topics belong to non-Brownian motion or anomalous and nonergodic diffusion. It will greatly benefit and promote the development of the particular research subject if the corresponding researchers can grasp the whole picture of the original scientific problems. The naive way is to read all the related literatures, which not only is an infeasible task, but it confuses and misunderstands the concepts in different disciplines.

Luckily, our research group is performing the interdisciplinary research, and we have good research collaborations with mathematicians, computational scientists, statistical physicists, biological physicists, and even chemists. This makes us to easily understand all their languages and their recognitions to the scientific problems. So the overarching goal of this book is to make the scientists from different research backgrounds easily reach the frontier of anomalous diffusion, rapidly master its big picture, and carry out research.

In Chapter 1, we introduce several common physical models and the stochastic processes to describe the anomalous and nonergodic diffusion in the natural world. Then we focus on the central part of the monograph: distribution of statistical observables. We analyze the probability density function (PDF) of the underlying processes for anomalous and nonergodic diffusions, and further derive the macroscopic equations governing the PDFs of the particle's position and the corresponding functionals. Besides, we extend our studies to more general statistical observables, such as the fractional moments, first passage time and first hitting time, and present their connections to mathematics and other applications.

One typical connection to mathematics is the Fokker-Planck equation governing the PDFs of particle's position. Recently, the fractional Fokker-Planck system with multiple internal states has been built. Here we provide the efficient numerical methods for the new model. To be specific, we first propose numerical scheme for the time fractional Fokker-Planck system with two internal states, including the first- and second-order scheme generated by convolution quadrature. And then, we consider the regularity and the fully discrete scheme constructed by L1 scheme and finite element method for the time-space fractional Fokker-Planck system. Also, the extensive numerical experiments are provided to validate the effectiveness of our schemes.

Chapter three is for the stochastic differential equation governing the probability density function of the statistical observables. Sometimes, to more practically explain phenomena and fit data in complex environments, it is necessary to pay attention to all the external stochastic disturbances. We establish the regularity of the equations, design the numerical schemes, perform the error analyses, and make some numerical experiments to verify the presented theoretical results.

The aim of this monograph is to make the scientists from different research backgrounds more easily access the topical scientific problem: anomalous and nonergodic diffusion. For this, we try to use the simple language well understood by the scholars from different scientific disciplines. We balance the intuitive understanding of the physical objects and strictly mathematical characterization by dealing with the same statistical observables from both physical and mathematical perspectives, build the bridge between stochastic processes and PDEs, and detect the difficulties of fractional PDEs from physics or mathematics.

CHAPTER 1

Statistical Observables

1.1 INTRODUCTION

Diffusion is a fundamental phenomenon that links the microscopic and macroscopic properties of physical systems. The study of diffusion phenomenon is originated with the Scottish botanist Robert Brown. In his 1827 fertilization study, he observed the irregular movement of floret powder particle suspended in water. The particle's motion comes neither from the current in the fluid nor from its gradual evaporation, but from the particles themselves. Later, Albert Einstein made remarkable contributions to the study of Brownian motion, and his 1905 paper [55] became the basis for probabilistic statistical mechanics. In the first part of his work, he used the theory established by Adolf Fick in 1855 to obtain the relationship between fluid viscosity and diffusion coefficient by balancing osmotic pressure and diffusion flow. In another part of the paper, Einstein considered a simple random walk with bounded and independent increments. Then he found that the probability density function (PDF) $p(x,t)$ of the particle's displacement satisfying the diffusion equation

$$\frac{\partial}{\partial t}p(x,t) = D\frac{\partial^2}{\partial x^2}p(x,t). \tag{1.1}$$

It was soon realized that the normal diffusion process represented only part of the diffusion processes. The anomalous diffusion

DOI: 10.1201/9781003279099-1

phenomena are also ubiquitous in the natural world, especially in numerous microscopic systems. Over the last two decades, much effort has been devoted to the study of anomalous diffusion phenomena [24, 76, 136], characterized by nonlinear time dependence of mean squared displacement (MSD)

$$\langle x^2(t) \rangle \simeq 2K_\gamma t^\gamma, \tag{1.2}$$

where $\gamma \neq 1$ is the anomalous diffusion exponent and K_γ is the anomalous diffusion coefficient with physical dimension $[\text{cm}^2/\text{s}^\gamma]$; it is called subdiffusion for $0 < \gamma < 1$ and superdiffusion for $\gamma > 1$.

1.1.1 Physical Models

The classical models describing Brownian motion correspond to the motion of a particle in a homogeneous, quiescent fluid, i.e., in a rather simple environment. Various anomalous diffusion phenomena are observed in complex environment, which promotes the physicists to formulate a roughest possible (minimal) model still reasonably describing these phenomena. The most common are continuous time random walk (CTRW) and (generalized) Langevin equation.

1.1.1.1 *Continuous time random walk*

In the pioneering work published in year 1965 [141], physicists Eliott W. Montroll and George H. Weiss introduced the concept of CTRW as a way to make the interevent-time continuous and fluctuating. It is characterized by some distribution associated with a stochastic process, giving an insight into the process activity. This distribution, called waiting time, permitted the description of both Debye (exponential) and, what is most significant, non-Debye (slowly-decaying) relaxations as well as normal and anomalous transport and diffusion [99, 136, 197]. Thus the model involves fundamental aspects of the stochastic world—a real, complex world.

The CTRW model contains two kinds of random variables, jump length and waiting time. The former describes the length of a given jump while the latter is the elapsing between two successive jumps. The joint PDF of jump length and waiting time is $\phi(x,t)$, from which the marginal jump length PDF

$$w(x) = \int_0^\infty \phi(x,t)dt \tag{1.3}$$

and the marginal waiting time PDF

$$\psi(t) = \int_{-\infty}^{\infty} \phi(x, t) dx \qquad (1.4)$$

can be obtained. Thus, $w(x)dx$ denotes the probability of a jump length in the interval $[x, x + dx]$ and $\psi(t)dt$ the probability of a waiting time in the interval $[t, t + dt]$. If the waiting time and jump length are independent random variables, it holds that $\phi(x, t) = w(x)\psi(t)$; otherwise, there is $\phi(x, t) = p(x|t)\psi(t)$ or $\phi(x, t) = p(t|x)w(x)$, which means that a certain jump length involves a time cost. Although the description of CTRW does not seem too complicated, it can describe both normal and anomalous diffusion and uncover the essential difference between them.

Now, let us see how to extract the macroscopic statistical information from this model. For the decoupled case with $\phi(x, t) = w(x)\psi(t)$, let $\gamma(x, t)$ be the incoming flow, i.e., the particles arriving at the point x per unit of time. The transport equation connecting the flows from the neighboring points can be written as

$$\gamma(x, t) = \int_{-\infty}^{\infty} \int_{0}^{t} w(x')\psi(t')\gamma(x - x', t - t')dx'dt' + p_0(x)\delta(t). \quad (1.5)$$

The integration term on the right-hand side denotes the flow $\gamma(x - x', t - t')$ which arrives at the point x at time t through a jump length x' and a waiting time t' within one step. The second term on the right-hand side assumes that the initial distribution of particles is $p_0(x)$. The next step is to relate the incoming flow $\gamma(x, t)$ to the current density of particles $p(x, t)$ for a given point at space x and time t:

$$p(x, t) = \int_{0}^{t} \Psi(t')\gamma(x, t - t')dt', \qquad (1.6)$$

where $\Psi(t)$ is the so-called survival probability, representing the probability that the waiting time on a site exceeds time t, i.e.,

$$\Psi(t) = \int_{t}^{\infty} \psi(t')dt' = 1 - \int_{0}^{t} \psi(t')dt'. \qquad (1.7)$$

The Laplace transform of $\Psi(t)$ can be obtained by using the property of Laplace transform of an integral [2] and reads

$$\hat{\Psi}(\lambda) = \frac{1 - \hat{\psi}(\lambda)}{\lambda}. \qquad (1.8)$$

The integration term on the right-hand side of Eq. (1.6) means that the incoming flow $\gamma(x, t - t')$ makes the last jump to position x and stays there before time t.

Considering the convolution form in Eqs. (1.5) and (1.6), we perform the Fourier-Laplace transform $(x \to k, t \to \lambda)$ on them. Thus, the incoming flow $\gamma(x, t)$ can be explicitly given in frequency domain by

$$\tilde{\hat{\gamma}}(k, \lambda) = \frac{\tilde{p}_0(k)\hat{\psi}(\lambda)}{1 - \tilde{w}(k)\hat{\psi}(\lambda)}. \tag{1.9}$$

Similarly, the Fourier-Laplace transform of Eq. (1.6) gives

$$\tilde{\hat{p}}(k, \lambda) = \tilde{w}(k)\hat{\Psi}(\lambda)\tilde{\hat{\gamma}}(k, \lambda) + \tilde{p}_0(k)\hat{\Psi}(\lambda). \tag{1.10}$$

Combining these two equations, we have

$$\tilde{\hat{p}}(k, \lambda) = \frac{\tilde{p}_0(k)\hat{\Psi}(\lambda)}{1 - \tilde{w}(k)\hat{\psi}(\lambda)}, \tag{1.11}$$

which is the well-known Montroll-Weiss equation [141]. It is the central result of the theory of CTRWs and is also valid for the higher dimensions. The initial condition is commonly assumed to be $p_0(x) = \delta(x)$. In this case, it holds that $\tilde{p}_0(k) = 1$.

Now we pay attention to the coupled case with $\phi(x, t) = p(x|t)\psi(t)$. The simplest way of coupling is

$$p(x|t) = \frac{1}{2}\delta(|x| - v_0 t), \tag{1.12}$$

which means that the particle moves on a straight line at a fixed velocity v_0 or $-v_0$ for a random time t. At the end of each excursion, the particle randomly chooses a new direction of its motion and moves for another random time with the same velocity v_0. The choice of direction has the equal probability $1/2$ to left or right. The durations of each excursion are independent and drawn from the same distribution $\psi(t)$. This model is named as Lévy walk. It is able to describe various regimes of stochastic transport, from normal diffusion to superdiffusion.

Similar to the derivations in uncoupled case, we now also present the transport equations for Lévy walk. Let $\gamma(x, t)$ be the frequency of velocity change at position x at time t, which satisfies

$$\gamma(x, t) = \int_{-\infty}^{\infty} \int_0^t \phi(x', t')\gamma(x - x', t - t')dx'dt' + p_0(x)\delta(t). \tag{1.13}$$

The last term in Eq. (1.13) means that the initial distribution of particles is $p_0(x)$. On the other hand, the equation connecting $\gamma(x, t)$ and the current density $p(x, t)$ is

$$p(x, t) = \int_{-\infty}^{\infty} \int_0^t \Phi(x', t')\gamma(x - x', t - t')dx'dt', \qquad (1.14)$$

where

$$\Phi(x', t') = \frac{1}{2}\delta(|x'| - v_0 t')\Psi(t'). \qquad (1.15)$$

The integral on the right-hand side of Eq. (1.14) shows that the particles from position $x - x'$ at time $t - t'$ undergo a single running event before time t. The length of this running event is $x' = \pm v_0 t'$. Equations (1.13) and (1.14) can be solved by using Fourier-Laplace transform, together with an additional technical complexity due to the coupling between time and space variables [97, 198]. We resolve it by using the shift property of Fourier-Laplace transforms and obtain:

$$\hat{\tilde{p}}(k, \lambda) = \frac{[\hat{\tilde{\Psi}}(\lambda + ikv_0) + \hat{\tilde{\Psi}}(\lambda - ikv_0)]\hat{p}_0(k)}{2 - [\hat{\tilde{\psi}}(\lambda + ikv_0) + \hat{\tilde{\psi}}(\lambda - ikv_0)]}, \qquad (1.16)$$

which is a generalization of the Montroll-Weiss equation (1.11) to the coupled case.

Lévy walk dynamics describe enhanced transport phenomena in many systems, such as diffusion in Josephson junctions [67], diffusion of atoms in optical lattices [17, 128], and blinking statistics of quantum dots [91]. Within the CTRW framework, the significant feature of Lévy walk is the underlying spatiotemporal coupling, which penalizes long jumps and leads to a finite MSD [197], while the uncoupled process, Lévy flight [136, 164], has divergent MSD.

1.1.1.2 Langevin equation

In the early 20th century, much important work was devoted to studying the irregular motion of microscopic particles in fluids, known as Brownian motion. Although the trajectory of Brownian motion is irregular, people can still find some hidden laws through statistical analysis, such as the MSD increases linearly with respect to time. Einstein's basic idea was to explain Brownian motion in terms of kinetic theory instead of thermodynamics, and the study of diffusion in Einstein's paper [55] and Smoluchowski's work [177] laid the foundation for the modern theory of Brownian motion. The further development of Brownian motion was that Langevin combined Newton's second law and wrote the motion equation of Brownian particle under the assumption that Brownian particle was subjected to two kinds of forces. Next we will briefly introduce the classical and generalized Langevin equations.

Consider the situation that a particle moves in a heat bath and collides with the surrounding water molecules. Assuming that the particle has a mass of $m = 1$, which is much larger than the water molecules in the heat bath, then according to Newton's second law, the trajectory of the particle can be described by the following stochastic differential equation [38, 105, 107]:

$$\ddot{x}(t) = -\gamma \dot{x}(t) + \xi(t), \tag{1.17}$$

where $x(t)$ and $\dot{x}(t) = v(t)$ are the position and velocity of the particle, respectively, γ is a frictional constant, and $\xi(t)$ is the Gaussian white noise with zero mean and correlation function

$$\langle \xi(t_1)\xi(t_2) \rangle = 2k_B \mathcal{T} \gamma \delta(t_1 - t_2), \tag{1.18}$$

where k_B is the Boltzmann constant and $\mathcal{T}$ the absolute temperature of the environment. The Eq. (1.18) is named as the fluctuation-dissipation theorem [102, 105]. It says that the effects of the surrounding molecules are generally two folds: the random driving force $\xi(t)$ and the frictional force $\gamma v(t)$. Since the two kinds of effects come from the same source (the surrounding molecules) so that they satisfy the quantitative relationship in Eq. (1.18). The stochastic equation (1.17) is called underdamped Langevin equation, as the acceleration term $\ddot{x}(t)$ is included.

Considering the property of linear additivity of Gaussian process, we find that the Browian motion described by Eq. (1.17) is a Gaussian process since the noise $\xi(t)$ is Gaussian. Therefore, the PDF of Brownian motion in Eq. (1.17) can be fully determined by its first and second moments. Corresponding to Eq. (1.17), the Langevin equation of the velocity process $v(t)$ is

$$\dot{v}(t) = -\gamma v(t) + \xi(t). \tag{1.19}$$

Assuming the initial condition is $x(0) = v(0) = 0$, based on Eq. (1.19), the velocity process $v(t)$ can be obtained by using the technique of Laplace transform:

$$v(t) = \int_0^t e^{-\gamma(t-t')} \xi(t') dt'. \tag{1.20}$$

Taking the ensemble average on both sides, we obtain the mean of velocity

$$\langle v(t) \rangle = 0, \tag{1.21}$$

which implies that the mean displacement is

$$\langle x(t) \rangle = x(0) + \int_0^t \langle v(t') \rangle dt' = 0. \tag{1.22}$$

The velocity correlation function can be obtained by using the correlation function of Gaussian white noise $\xi(t)$ in Eq. (1.18), i.e.,

$$\langle v(t_1)v(t_2) \rangle = k_B \mathcal{T} \left(e^{-\gamma|t_1-t_2|} - e^{-\gamma(t_1+t_2)} \right). \tag{1.23}$$

For large t_1 and t_2 so that $\gamma t_1 \gg 1$, $\gamma t_2 \gg 1$, the asymptotic expression of velocity correlation function is

$$\langle v(t_1)v(t_2) \rangle \simeq k_B \mathcal{T} e^{-\gamma|t_1-t_2|}, \tag{1.24}$$

which is independent on the initial velocity v_0, but only depends on the time difference $|t_1 - t_2|$. This means that the velocity process is asymptotically stationary. For large time t, the MSD of the Brownian motion described by Langevin equation (1.17) is

$$\begin{aligned}
\langle x^2(t) \rangle &= \int_0^t \int_0^t \langle v(t_1)v(t_2) \rangle dt_2 dt_1 \\
&= 2 \int_0^t \int_0^{t_1} \langle v(t_1)v(t_2) \rangle dt_2 dt_1 \\
&\simeq \frac{2k_B \mathcal{T}}{\gamma} t = 2Dt,
\end{aligned} \tag{1.25}$$

where D is called diffusivity, and the equality $D = k_B \mathcal{T}/\gamma$ is named as the Einstein relation [105]. This shows that the Brownian motion in Eq. (1.17) exhibits the normal diffusion. Combining the mean and variance of Brownian motion described by Eqs. (1.22) and (1.25), we find the PDF of Brownian motion is

$$p(x,t) = \frac{1}{\sqrt{4\pi Dt}} \exp\left(-\frac{x^2}{4Dt}\right). \tag{1.26}$$

If we assume that the friction is quite large so that the velocity reaches a steady state in a short time, then we can omit the acceleration term $\ddot{x}(t)$ in Eq. (1.17), and obtain the overdamped Langevin equation [107]:

$$\dot{x}(t) = \frac{1}{\gamma}\xi(t). \tag{1.27}$$

Similar to the analyses on the underdamped Langevin equation, we find the mean displacement is still zero and the MSD is

$$\langle x^2(t)\rangle = \frac{1}{\gamma^2}\int_0^t\int_0^t\langle R(t_1)R(t_2)\rangle dt_2 dt_1 = \frac{2k_B\mathcal{T}t}{\gamma}, \tag{1.28}$$

the same as the one in Eq. (1.25) for the underdamped case.

The Langevin equation (1.17) describes the normal diffusion process. By changing the formation of frictional force, however, its variant can describe anomalous diffusion processes. For example, the generalized Langevin equation describing a free particle with mass $m = 1$ and driven by an internal noise $\rho(t)$ reads [38]

$$\ddot{x}(t) = -\int_0^t K(t-\tau)\dot{x}(\tau)d\tau + \rho(t), \tag{1.29}$$

where $x(t)$ denotes the particle displacement, $\dot{x}(t) = v(t)$ is the particle velocity, $K(t)$ is the friction memory kernel, and $\rho(t)$ is the random driving force subject to the conditions $\langle\rho(t)\rangle = 0$ and satisfies the fluctuation-dissipation theorem [105]

$$\langle\rho(t_1)\rho(t_2)\rangle = k_B\mathcal{T}K(t_1 - t_2), \tag{1.30}$$

which is a generalization of Eq. (1.18). This is why $\rho(t)$ is named as internal noise. The corresponding overdamped generalized Langevin equation without acceleration term is written as

$$0 = -\int_0^t K(t-\tau)\dot{x}(\tau)d\tau + \rho(t). \tag{1.31}$$

There are two common kinds of generalized Langevin equation: fractional Langevin equation [38, 48, 102, 125] and tempered fractional Langevin equation [36, 37, 140]. The corresponding noises are the fractional Gaussian noise and tempered fractional Gaussian noise, respectively, as the generalization of white Gaussian noise. The processes described by (tempered) fractional Langevin equation are still Gaussian due to the linearity of the generalized Langevin equation. However, they present anomalous diffusion behavior distinguishing from Brownian motion and classical Langevin equation.

1.1.2 Stochastic Processes

In this section, we introduce several common stochastic processes when describing anomalous diffusion phenomena, which are Lévy processes, subordinator, and time-changed processes.

1.1.2.1 Lévy process

A stochastic process $X(t)$ for $t \geq 0$ is a Lévy process if it satisfies the following conditions [12]:

1. $X(0) = 0$ almost surely, i.e., for each of its different realizations.

2. $X(t)$ has independent increments, i.e., for all $n \geq 2$ and for each partition $0 \leq t_0 < t_1 \cdots < t_n \leq t$, the random variables $X(t_j) - X(t_{j-1})$, $j = 1, 2, \cdots, n$ are independent.

3. $X(t)$ has stationary increments, i.e., for all $0 \leq t_1 < t_2 \leq t$, the random variables $X(t_2) - X(t_1)$ has the same distribution with $X(t_2 - t_1)$.

4. The trajectories of $X(t)$ are càdlàg, i.e., right continuous with left limits.

If one restricts the conditions (2) and (4) with Gaussian distributed increments and continuous trajectories, respectively, one recovers the ordinary Brownian motion.

Another important property of Lévy process is infinitely divisible. It says that Lévy process $X(t)$ can be expressed as a sum of different independent identical distribution (i.i.d.) random variables Y_j, $j = 1, 2, \cdots, n$ for any n:

$$X(t) = \sum_{j=1}^{n} Y_j. \tag{1.32}$$

Let the characteristic functions of $X(t)$ and Y_j be $\phi_X(k)$ and $\phi_Y(k)$, respectively, i.e.,

$$\phi_X(k) = \langle e^{ikX(t)} \rangle, \quad \phi_Y(k) = \langle e^{ikY_j} \rangle. \tag{1.33}$$

Then, their characteristic functions are related by

$$\phi_X(k) = \left\langle e^{ik\sum_{j=1}^n Y_j} \right\rangle = \prod_{j=1}^n \langle e^{ikY_j} \rangle = [\phi_Y(k)]^n. \tag{1.34}$$

Due to the infinite divisibility, the characteristic function of a Lévy process can be written as

$$\phi_X(k,t) = \langle e^{ikX(t)} \rangle = e^{G(k)t}, \tag{1.35}$$

which is what the Lévy-Khintchine formula says. Especially, $G(k)$ is named as Lévy symbol, and it has a specific form [12]

$$G(k) = ikb - \frac{1}{2}k^2 a + \int_{\mathbb{R}\backslash\{0\}} \left[e^{iky} - 1 - iky\chi_{\{|y|<1\}} \right] \nu(dy), \tag{1.36}$$

where $a > 0$ and b are constants, χ_I is the indicator function of the set I, and ν is a finite Lévy measure on $\mathbb{R}\backslash\{0\}$ satisfies

$$\int_{\mathbb{R}\backslash\{0\}} (1 \wedge y^2)\nu(dy) < \infty. \tag{1.37}$$

Thus, any Lévy process $X(t)$ can be uniquely characterized by the triplet (b, a, ν).

The most typical examples of Lévy process are Brownian motion and Lévy stable process. Brownian motion is also named as Wiener process. It has continuous sample paths and Gaussian PDF. The Lévy symbol $G(k)$ of Brownian motion only contains the former two terms in Eq. (1.36). The last term in Eq. (1.36) characterizes the jump feature of Lévy process. For standard Brownian motion $B(t)$ with variance t, its characteristic function is given by

$$\langle e^{ik \cdot B(t)} \rangle = e^{-\frac{1}{2}tk^2}. \tag{1.38}$$

Brownian motion is the only Lévy process possessing continuous sample paths. By contrast, the Lévy stable process is a stable random variable at any instant. Of particular interest is the rotationally invariant case, where the Lévy symbol is

$$G(k) = -\sigma^\alpha |k|^\alpha, \tag{1.39}$$

where $0 < \alpha \leq 2$ is the index of stability and the scale parameter $\sigma > 0$. One noteworthy feature of the Lévy stable process is the self-similarity. In general, a stochastic process $X(t)$ is self-similar with Hurst index $H > 0$ if the two processes $X(at)$ and $a^H X(t)$ have the same finite-dimensional distributions for any $a \geq 0$. Within this definition, it can be verified that the rotationally invariant Lévy stable process is self-similar with Hurst index $H = 1/\alpha$. For the special case of Brownian motion ($\alpha = 2$), it is self-similar with Hurst index $H = 1/2$. More self-similar processes can be found in [56].

1.1.2.2 Subordinator

A subordinator is a one-dimensional Lévy process that is a.s. non-decreasing [12]. Thus, if $X(t)$ is a subordinator for $t \geq 0$, the following properties hold:

1. For all $t \geq 0$, $X(t) \geq 0$.

2. For all $t_1 \leq t_2$, $X(t_1) \leq X(t_2)$.

Since a subordinator is a Lévy process, it should be uniquely characterized by a triplet (b, a, ν) in Eq. (1.36). The strictly mathematical definition of a subordinator is that its Lévy symbol takes the form

$$G(k) = ikb + \int_0^\infty (e^{iky} - 1)\nu(dy), \qquad (1.40)$$

where $b \geq 0$ and the Lévy measure ν satisfies the additional conditions

$$\nu(-\infty, 0) = 0 \quad \text{and} \quad \int_0^\infty (y \wedge 1)\nu(dy) < \infty. \qquad (1.41)$$

The pair (b, ν) is named as the characteristics of the subordinator $X(t)$. Since $X(t) \geq 0$, the characteristic function is generally given through the Laplace transform:

$$\langle e^{-\lambda X(t)} \rangle = e^{-t\Phi(\lambda)}, \qquad (1.42)$$

where

$$\Phi(\lambda) = -G(i\lambda) = b\lambda + \int_0^\infty (1 - e^{-\lambda y})\nu(dy) \qquad (1.43)$$

for any $\lambda > 0$. Here, $\Phi(\lambda)$ is often called Laplace exponent.

The most common subordinator is Lévy stable subordinator. A subordinator $X(t)$ is Lévy stable if it has the characteristic pair:

$$b = 0, \qquad \nu(dy) = \frac{\alpha}{\Gamma(1-\alpha)} y^{-1-\alpha} dy, \qquad (1.44)$$

where $0 < \alpha < 1$. Substituting the characteristic pair into Eq. (1.43), we obtain the Laplace exponent:

$$\begin{aligned}
\Phi(\lambda) &= \frac{\alpha}{\Gamma(1-\alpha)} \int_0^\infty (1 - e^{-\lambda y}) y^{-1-\alpha} dy \\
&= \frac{\lambda}{\Gamma(1-\alpha)} \int_0^\infty e^{-\lambda y} y^{-\alpha} dy \\
&= \lambda^\alpha,
\end{aligned} \qquad (1.45)$$

where we used the integration by parts in the second line and the result of Laplace transform in the last line. Another example of subordinator is the tempered Lévy stable subordinator. The characteristic pair of a tempered Lévy stable subordinator $X(t)$ is

$$b = 0, \qquad \nu(dy) = \frac{\alpha}{\Gamma(1-\alpha)} e^{-\mu y} y^{-1-\alpha} dy, \qquad (1.46)$$

where $0 < \alpha < 1$ and $\mu > 0$. Substituting the characteristic pair into Eq. (1.43), we obtain the Laplace exponent:

$$\begin{aligned}
\Phi(\lambda) &= \frac{\alpha}{\Gamma(1-\alpha)} \int_0^\infty (1 - e^{-\lambda y}) e^{-\mu y} y^{-1-\alpha} dy \\
&= \frac{1}{\Gamma(1-\alpha)} \int_0^\infty \left((\lambda + \mu) e^{-(\lambda+\mu)y} - \mu e^{-\mu y} \right) y^{-\alpha} dy \\
&= (\lambda + \mu)^\alpha - \lambda^\alpha,
\end{aligned} \qquad (1.47)$$

where we used the integration by parts in the second line and the result of Laplace transform in the last line.

1.1.2.3 *Time-changed process*

The subordinator is usually used to build a time-changed process. In the framework of Langevin equation, the time-changed process can be expressed as

$$\dot{x}(s) = \xi(s), \qquad \dot{t}(s) = \eta(s), \qquad (1.48)$$

where $x(s)$ and $t(s)$ denote the position and time in physical space. The noises $\xi(s)$ and $\eta(s)$ are responsible for the stochastic character of the process. Since the physical time t becomes a process here, the process $x(t)$ concerned is named as time-changed process. To guarantee the positivity and non-decreasing of physical time t, it is usually taken as a subordinator. For convenience, let $t(s)$ be the Lévy stable subordinator with $\alpha < 1$ and $x(s)$ be the standard Brownian motion for further analyses.

The strict definition of the time-changed process $x(t)$ in Eq. (1.48) needs the idea of inverse subordinator $s(t)$, which is defined as

$$s(t) = \inf_{s>0}\{s : t(s) > t\}, \tag{1.49}$$

being the first passage time of the subordinator $t(s)$. Therefore, the time-changed process $x(t)$ is actually denoted as $x(t) := x(s(t))$. Now, let us study the PDF of the time-changed process $x(t)$. Its PDF $p(x,t)$ can be denoted by

$$
\begin{aligned}
p(x,t) &= \langle \delta(x - x(s(t))) \rangle \\
&= \int_0^\infty \langle \delta(x - x(s))\delta(s - s(t)) \rangle ds \\
&= \int_0^\infty \langle \delta(x - x(s)) \rangle \langle \delta(s - s(t)) \rangle ds \\
&= \int_0^\infty p_0(x,s)h(s,t)ds,
\end{aligned}
\tag{1.50}
$$

where $p_0(x,s)$ is the PDF of the original process $x(s)$ and $h(s,t)$ is the PDF of the inverse subordinator $s(t)$. We have utilized the property of δ function in the second line, the independence between processes $x(s)$, $t(s)$ in the third line, and the definition of a PDF as a δ function in the last line of Eq. (1.50). The PDF $p_0(x,s)$ of the original process $x(s)$ is assumed to be known. For standard Brownian motion, it is

$$p_0(x,s) = \frac{1}{\sqrt{2\pi t}} \exp\left(-\frac{x^2}{2t}\right). \tag{1.51}$$

Thus, the key of obtaining the PDF $p(x,t)$ of the time-changed process $x(s(t))$ is the PDF $h(s,t)$ of the inverse subordinator $s(t)$.

Similarly, the two-point joint PDF of $x(s(t))$ can be written as

$$p(x_2, t_2; x_2, t_1) = \langle \delta(x_2 - x(s(t_2)))\delta(x_1 - x(s(t_1)))\rangle$$

$$= \int_0^\infty \int_0^\infty \langle \delta(x_2 - x(s_2))\delta(x_1 - x(s_1))$$

$$\times \delta(s_2 - s(t_2))\delta(s_1 - s(t_1))\rangle ds_1 ds_2$$

$$= \int_0^\infty \int_0^\infty p_0(x_2, s_2; x_1, s_1)h(s_2, t_2; s_1, t_1)ds_1 ds_2,$$

$$(1.52)$$

where $p_0(x_2, s_2; x_1, s_1)$ and $h(s_2, t_2; s_1, t_1)$ are the two-point joint PDFs of the original process $x(s)$ and the inverse subordinator $s(t)$, respectively. Therefore, $h(s_2, t_2; s_1, t_1)$ is the premise of obtaining $p(x_2, t_2; x_2, t_1)$.

Now we turn to the discussions of the PDFs of the inverse subordinator $s(t)$. Since $s(t)$ is the inverse process of the subordinator $t(s)$, as defined in Eq. (1.49), they satisfy the following relations

$$\langle \Theta(s - s(t))\rangle = 1 - \langle \Theta(t - t(s))\rangle, \tag{1.53}$$

and

$$\langle \Theta(s_2 - s(t_2))\Theta(s_1 - s(t_1))\rangle = 1 - \langle \Theta(t_2 - t(s_2))\rangle$$
$$- \langle \Theta(t_1 - t(s_1))\rangle + \langle \Theta(t_2 - t(s_2))\Theta(t_1 - t(s_1))\rangle, \tag{1.54}$$

where $\Theta(x)$ is the Heaviside step function satisfying

$$\Theta(x) = \begin{cases} 1, & \text{for } x > 0, \\ 0, & \text{for } x < 0, \\ \frac{1}{2}, & \text{for } x = 0. \end{cases} \tag{1.55}$$

Note that the PDFs of inverse subordinator $s(t)$ can be expressed as

$$h(s, t) = \langle \delta(s - s(t))\rangle, \tag{1.56}$$

$$h(s_2, t_2; s_1, t_1) = \langle \delta(s_2 - s(t_2))\delta(s_1 - s(t_1))\rangle, \tag{1.57}$$

which are just the derivatives with respect to s on Eq. (1.53) and s_1, s_2 on Eq. (1.54). Thus, we obtain

$$h(s, t) = -\frac{\partial}{\partial s}\langle \Theta(t - t(s))\rangle, \tag{1.58}$$

$$h(s_2, t_2; s_1, t_1) = \frac{\partial}{\partial s_1} \frac{\partial}{\partial s_2} \langle \Theta(t_2 - t(s_2))\Theta(t_1 - t(s_1)))\rangle. \qquad (1.59)$$

It can be seen that the ensemble averages on the right-hand sides of Eqs. (1.58) and (1.59) are over the realizations of the subordinator $t(s)$, which can be explicitly calculated since it is a Lévy process. Let the PDFs of the subordinator $t(s)$ be $Z(t, s)$ and $Z(t_2, s_2; t_1, s_1)$, respectively. The Laplace transform $(t \to \lambda)$ of $Z(t, s)$ is

$$\hat{Z}(\lambda, s) = \langle e^{-\lambda t(s)} \rangle = e^{-s\lambda^\alpha}, \qquad (1.60)$$

for $0 < \alpha < 1$. The expression in real domain is

$$Z(t, s) = \frac{1}{s^{1/\alpha}} L_\alpha \left(\frac{t}{s^{1/\alpha}} \right), \qquad (1.61)$$

where $L_\alpha(t)$ represents the one-sided Lévy-stable distribution with Laplace transform $\mathcal{L}\{L_\alpha(t)\} = e^{-\lambda^\alpha t}$. The corresponding two-point PDF $Z(t_2, s_2; t_1, s_1)$ will be obtained by using the property of independence and stationarity of increments. In detail, the double Laplace transform $(t_1 \to \lambda_1, t_2 \to \lambda_2)$ of the PDF $Z(t_2, s_2; t_1, s_1)$ is

$$\hat{Z}(\lambda_2, s_2; \lambda_1, s_1) = \left\langle e^{-\lambda_2 t(s_2) - \lambda_1 t(s_1)} \right\rangle$$

$$= \Theta(s_2 - s_1) \left\langle \exp\left(-\lambda_2 \int_{s_1}^{s_2} ds' \, \eta(s') - (\lambda_1 + \lambda_2) \int_0^{s_1} ds' \, \eta(s') \right) \right\rangle$$

$$+ \Theta(s_1 - s_2) \left\langle \exp\left(-\lambda_1 \int_{s_1}^{s_2} ds' \, \eta(s') - (\lambda_1 + \lambda_2) \int_0^{s_2} ds' \, \eta(s') \right) \right\rangle. \qquad (1.62)$$

The ensemble average can be split into two parts due to the independence of increments of Lévy process. Thus we have

$$\hat{Z}(\lambda_2, s_2; \lambda_1, s_1) = \Theta(s_2 - s_1) e^{-s_1(\lambda_1 + \lambda_2)^\alpha} e^{-(s_2 - s_1)\lambda_2^\alpha}$$

$$+ \Theta(s_1 - s_2) e^{-s_2(\lambda_1 + \lambda_2)^\alpha} e^{-(s_1 - s_2)\lambda_1^\alpha}. \qquad (1.63)$$

Combining Eqs. (1.58) and (1.60) yields the single-point PDF of inverse subordinator $s(t)$:

$$\hat{h}(s, \lambda) = -\frac{\partial}{\partial s} \left\langle \frac{1}{\lambda} e^{-\lambda t(s)} \right\rangle = \lambda^{\alpha-1} e^{-s\lambda^\alpha}. \qquad (1.64)$$

Its inverse Laplace transform is also the Lévy distribution: [16]

$$h(s, t) = \frac{1}{\alpha} \frac{t}{s^{1+1/\alpha}} L_\alpha \left(\frac{t}{s^{1/\alpha}} \right). \qquad (1.65)$$

The normalization of PDF $h(s,t)$ can be verified by integrating with respect to s in Eq. (1.64), i.e.,

$$\int_0^\infty \hat{h}(s,\lambda)ds = \frac{1}{\lambda}, \tag{1.66}$$

the inverse Laplace transform of which is 1. On the other hand, Eq. (1.64) can be rewritten as

$$\lambda \hat{h}(s,\lambda) = -\frac{\partial}{\partial s}e^{-s\lambda^\alpha}. \tag{1.67}$$

Noting that the initial condition of the PDF of the inverse subordinator $s(t)$ is $h(s,0) = \delta(s) = 0$ for $s > 0$, performing the inverse Laplace transform on Eq. (1.67) gives a fractional evolution equation:

$$\frac{\partial}{\partial t}h(s,t) = -D_t^{1-\alpha}\frac{\partial}{\partial s}h(s,t), \tag{1.68}$$

where the operator $D_t^{1-\alpha}$ denotes the Riemann-Liouville fractional differential operator.

Similarly, we combine Eqs. (1.59) and (1.63) with some calculations, and obtain the two-point PDF $h(s_2,t_2;s_1,t_1)$ in Laplace space

$$\begin{aligned}
\hat{h}(s_2,\lambda_2;s_1,\lambda_1) &= \frac{\partial}{\partial s_1}\frac{\partial}{\partial s_2}\frac{1}{\lambda_1\lambda_2}\hat{p}(\lambda_2,s_2;\lambda_1,s_1) \\
&= \delta(s_2 - s_1)\frac{\lambda_1^\alpha + \lambda_2^\alpha - (\lambda_1 + \lambda_2)^\alpha}{\lambda_1\lambda_2}e^{-s_1(\lambda_1+\lambda_2)^\alpha} \\
&\quad + \Theta(s_2 - s_1)\frac{\lambda_2^\alpha[(\lambda_1 + \lambda_2)^\alpha - \lambda_2^\alpha]}{\lambda_1\lambda_2} \\
&\quad \times e^{-s_1(\lambda_1+\lambda_2)^\alpha}e^{-(s_2-s_1)\lambda_2^\alpha} \\
&\quad + \Theta(s_1 - s_2)\frac{\lambda_1^\alpha[(\lambda_1 + \lambda_2)^\alpha - \lambda_1^\alpha]}{\lambda_1\lambda_2} \\
&\quad \times e^{-s_2(\lambda_1+\lambda_2)^\alpha}e^{-(s_1-s_2)\lambda_1^\alpha}.
\end{aligned} \tag{1.69}$$

The normalization of the two-point PDF can be verified by integrating with respect to s_1 and s_2 in Eq. (1.69), i.e.,

$$\int_0^\infty\int_0^\infty \hat{h}(s_2,\lambda_2;s_1,\lambda_1)ds_2ds_1 = \frac{1}{\lambda_1\lambda_2}, \tag{1.70}$$

since the inverse Laplace transform of $\frac{1}{\lambda_1\lambda_2}$ is 1.

With the two-point PDF $h(s_2, t_2; s_1, t_1)$ in Eq. (1.69), we can calculate the correlation function of the time-changed process. Multiplying x_2 and x_1 on both sides of Eq. (1.52), and making the integral with respect to x_2 and x_1, we obtain

$$\langle x(t_2)x(t_1) \rangle = \int_0^\infty \int_0^\infty \langle x(s_2)x(s_1) \rangle h(s_2, t_2; s_1, t_1) ds_1 ds_2, \quad (1.71)$$

where $\langle x(s_2)x(s_1) \rangle$ is the correlation function of original process $x(s)$, satisfying

$$\langle x(s_2)x(s_1) \rangle = \Theta(s_2 - s_1)s_1 + \Theta(s_1 - s_2)s_2. \quad (1.72)$$

Performing the double Laplace transform ($t_1 \to \lambda_1, t_2 \to \lambda_2$) on Eq. (1.71) and substituting Eqs. (1.69) and (1.72) into it, we have

$$
\begin{aligned}
\mathcal{L}\{\langle x(t_2)x(t_1) \rangle\} &= \frac{(\lambda_1 + \lambda_2)^\alpha}{\lambda_1 \lambda_2} \int_0^\infty s e^{-s(\lambda_1 + \lambda_2)^\alpha} ds \\
&= \frac{1}{(\lambda_1 + \lambda_2)^\alpha \lambda_1 \lambda_2}.
\end{aligned}
\quad (1.73)
$$

Its inverse Laplace transform is

$$\langle x(t_2)x(t_1) \rangle = \frac{1}{\Gamma(\alpha + 1)} \left[\Theta(t_2 - t_1)t_1^\alpha + \Theta(t_1 - t_2)t_2^\alpha \right]. \quad (1.74)$$

Taking $t_2 = t_1 = t$, we find the time-changed process $x(s(t))$ exhibits subdiffusion:

$$\langle x^2(t) \rangle = \frac{t^\alpha}{\Gamma(\alpha + 1)} \quad (1.75)$$

for $\alpha < 1$.

1.2 POSITION

The position of a particle at a given time is an intuitive statistical quantity of interest. Since the position is not a deterministic function over time t, people are always studying the PDF $p(x, t)$ of particles. For pure Brownian motion, the PDF is Gaussian:

$$p(x, t) = \frac{1}{\sqrt{4\pi Dt}} \exp\left(-\frac{x^2}{4Dt} \right), \quad (1.76)$$

where D is the diffusivity of particles. However, the PDF of an anomalous diffusion process is hard to be explicitly expressed like

Eq. (1.76). Even for the common anomalous diffusion processes, such as subdiffusive CTRW and Lévy flight, their PDFs are usually represented in frequency domain, i.e., $\tilde{\hat{p}}(k, \lambda)$.

The expressions $p(x, t)$ of the common anomalous diffusion processes in real domain are not easy to be obtained. In fact, there is another way to describe the PDFs of different processes, which is to build the governing equations of the PDFs. These equations are named as Fokker-Planck equation. Once obtaining the Fokker-Planck equation, we can theoretically analyze the properties of the solution $p(x, t)$ in the equations, and numerically compute the solution $p(x, t)$ and its evolution with respect to time by using the discretization methods, such as finite element method, finite difference method, spectral method, etc.

1.2.1 Probability Density Function

To obtain the expression $p(x, t)$ in real domain for common anomalous diffusion processes, we will seek help from a powerful but complex tool—Fox H-function $H_{p,q}^{m,n}(z)$. Its complexity comes from its definition [129, 130, 168, 169]:

$$
\begin{aligned}
H_{p,q}^{m,n}(z) &= H_{p,q}^{m,n}\left[z \, \middle| \, \begin{matrix} (a_p, A_p) \\ (b_q, B_q) \end{matrix} \right] \\
&= H_{p,q}^{m,n}\left[z \, \middle| \, \begin{matrix} (a_1, A_1), (a_2, A_2), \cdots, (a_p, A_p) \\ (b_1, B_1), (b_2, B_2), \cdots, (b_q, B_q) \end{matrix} \right] \quad (1.77) \\
&= \frac{1}{2\pi i} \int_L \chi(s) z^s \, ds
\end{aligned}
$$

with

$$
\chi(s) = \frac{\prod_{j=1}^m \Gamma(b_j - B_j s) \prod_{j=1}^n \Gamma(1 - a_j + A_j s)}{\prod_{j=m+1}^q \Gamma(1 - b_j + B_j s) \prod_{j=n+1}^p \Gamma(a_j - A_j s)}. \quad (1.78)
$$

With the help of Fox H-function, the PDF of the subdiffusive CTRW can be expressed as [136]

$$
p(x, t) = \frac{1}{\sqrt{4\pi t^\alpha}} H_{1,2}^{2,0}\left[\frac{x^2}{4t^\alpha} \, \middle| \, \begin{matrix} (1 - \alpha/2, \alpha) \\ (0, 1), (1/2, 1) \end{matrix} \right]. \quad (1.79)
$$

The power of the Fox H-function is embodied by its operation. The Fox H-function is closed with respect to Laplace transform and

Fourier transform. Besides, the asymptotic behaviors ($x \to 0$ and $x \to \infty$) of the Fox H-function are also known. By employing the standard theorems of the Fox H-function, we obtain the asymptotic stretched Gaussian behavior from Eq. (1.79)

$$p(x,t) \simeq \frac{1}{\sqrt{4\pi t^\alpha}} \sqrt{\frac{1}{2-\alpha} \left(\frac{2}{\alpha}\right)^{(1-\alpha)/(2-\alpha)}} \left(\frac{|x|}{\sqrt{t^\alpha}}\right)^{-(1-\alpha)/(2-\alpha)}$$
$$\times \exp\left(-\frac{2-\alpha}{2}\left(\frac{\alpha}{2}\right)^{\alpha/(2-\alpha)}\left[\frac{|x|}{\sqrt{t^\alpha}}\right]^{1/(1-\alpha/2)}\right)$$

(1.80)

for $x \gg \sqrt{t^\alpha}$. While for Lévy flight, the corresponding PDF is a Lévy distribution:

$$p(x,t) \simeq L_{\beta,0}\left(\frac{|x|}{t^{1/\beta}}\right),$$

(1.81)

where $L_{\beta,0}(z)$ represents the symmetric stable Lévy distribution with exponent β. It can be also expressed through Fox H function as [136]

$$p(x,t) = \frac{1}{\sqrt{4\pi t^\beta}} H^{2,0}_{1,2}\left[\frac{x^2}{4t^\beta} \middle| \begin{array}{c} (1-\beta/2,\beta) \\ (0,1),(1/2,1) \end{array}\right],$$

(1.82)

and this PDF has a power law asymptotics

$$p(x,t) \simeq \frac{t}{|x|^{1+\beta}}.$$

(1.83)

Compared with the PDFs of subdiffusive CTRW and Lévy flight, that of Lévy walk becomes more difficult due to the coupling between waiting time and jump length. Even the PDF of Lévy walk does not have a uniform expression for different α. When $0 < \alpha < 1$, the mean running time diverges and this model has a ballistic scaling $x \sim t$. In this case, the shape of PDF can be obtained by using a technical method similar to that proposed by [71]. This method assumes that the model has the ballistic scaling form:

$$p(x,t) = \frac{1}{t} F\left(\frac{x}{t}\right).$$

(1.84)

Then the problem is turned into the expression of scaling function $F(z)$. On the other hand, the analyses on the PDF are often performed in Fourier-Laplace space, which implies that the PDF can also be expressed as

$$\tilde{p}(k, \lambda) = \frac{1}{\lambda} f\left(\frac{ik}{\lambda}\right), \qquad (1.85)$$

where the scaling function f can be obtained from every specific $\tilde{p}(k, \lambda)$ within the ballistic scaling. Further, the Sokhotsky-Weierstrass theorem shows the relation between two scaling functions:

$$F(z) = -\frac{1}{\pi z} \lim_{\epsilon \to 0} \operatorname{Im} f\left(-\frac{1}{z + i\epsilon}\right). \qquad (1.86)$$

This formula brings us the convenience of avoiding to calculate the inverse Fourier-Laplace transforms on $\tilde{p}(k, \lambda)$.

Taking the standard Lévy walk with a constant velocity $v_0 = 1$ as an example, the PDF in Eq. (1.16) shows that the scaling function is

$$f(\zeta) = \frac{(1 - \zeta)^{\alpha-1} + (1 + \zeta)^{\alpha-1}}{(1 - \zeta)^{\alpha} + (1 + \zeta)^{\alpha}}. \qquad (1.87)$$

Then by using Eq. (1.86), we find the shape of the scaling function:

$$F(z) = \frac{\sin(\pi\alpha)}{\pi} \frac{|z - 1|^{\alpha}|z + 1|^{\alpha-1} + |z + 1|^{\alpha}|z - 1|^{\alpha-1}}{|z - 1|^{2\alpha} + |z + 1|^{2\alpha} + 2|z - 1|^{\alpha}|z + 1|^{\alpha}\cos(\pi\alpha)}, \qquad (1.88)$$

which is the Lamperti distribution [106]. Since Lévy walk has the finite velocity, which is $v_0 = 1$ in this example, it holds that $|z| = |x|/t \leq 1$. This yields a significance difference with Eqs. (1.80) and (1.83), where the position x of particle can be infinite.

For the standard Lévy walk with $1 < \alpha < 2$, the situation becomes more complex than the case of $0 < \alpha < 1$. The PDF here even cannot be expressed by a single formula like Eq. (1.88). The Lévy walk with $1 < \alpha < 2$ is a typical model presenting multiscale phenomenon. The PDF shows different shapes between its central part and tail part, where the central part denotes the region around $|x| \sim t^{1/\alpha}$ and tail part denotes the region around $|x| \sim t$. The two kinds of scaling relations coexist in the propagator of Lévy walk. They both play important roles in deciding the macroscopic quantities of Lévy walk and none of them can be ignored.

Similar to Lévy flight, the central part of the PDF of Lévy walk is also subjected to the generalized central limit theorem and given by the symmetric Lévy distribution

$$p_{\text{cen}}(x, t) \simeq \frac{1}{(K_\alpha t)^{1/\alpha}} L_{\alpha,0}\left(\frac{x}{(K_\alpha t)^{1/\alpha}}\right). \qquad (1.89)$$

While the tail part of the PDF of Lévy walk scales as $x \sim t$ due to the constant velocity. To describe the scaling relation $x \sim t$ at the outmost fronts of the propagator, a scaled ballistic variable $\xi = x/t$ was introduced in [152, 153], where the PDF of this variable is defined as

$$\mathcal{I}(\xi) = \lim_{t\to\infty} t^\alpha p_{\text{tail}}(x/t, t). \qquad (1.90)$$

This function is non-normalizable, i.e.,

$$\int_{-\infty}^{\infty} \mathcal{I}(\xi) d\xi = \infty, \qquad (1.91)$$

due to the power law singularity in the limit $\xi \to 0$, $\mathcal{I}(\xi) \propto |\xi|^{-1-\alpha}$. Therefore, $\mathcal{I}(\xi)$ is named as infinite density. The infinite density is non-normalizable, the concept of which was thoroughly investigated in mathematics issues [1, 173], and has been applied to physics successfully. Here, it is aimed to characterize the ballistic scaling $(x \sim t)$ of Lévy walk. In addition, the infinite density is usually discussed together with infinite-ergodic theory, for example, the Brownian motion in a logarithmic potential [7] and the Langevin system with multiplicative noise [110, 185]. The infinite density is still valuable in spite of its singularity. In [152, 153], the explicit expression of an infinite density is given as

$$\mathcal{I}(\xi) = B\left[\frac{\alpha \mathcal{Q}_\alpha(|\xi|)}{|\xi|^{1+\alpha}} - \frac{(\alpha - 1)\mathcal{Q}_{\alpha-1}(|\xi|)}{|\xi|^\alpha}\right], \qquad (1.92)$$

where

$$\mathcal{Q}_\alpha(\xi) = \int_{|\xi|}^{\infty} dv v^\alpha h(v). \qquad (1.93)$$

Here, the $h(v)$ is the velocity distribution, reducing to $(\delta(v - v_0) + \delta(v + v_0))/2$ for the case of two-point velocity $\pm v_0$. The two kinds of scaling relations meet in the intermediate region where both functions scale as

$$L_{\alpha,0}\left(\frac{x}{t^{1/\alpha}}\right) \simeq \mathcal{I}\left(\frac{x}{t}\right) \simeq x^{-1-\alpha}. \qquad (1.94)$$

1.2.2 Fokker-Planck Equation

Since the anomalous diffusion processes can be modeled from CTRW or Langevin equation, the corresponding Fokker-Planck equations can be derived under these two different frameworks. The differences in derivations are just technical. If the two models describe the same anomalous process, the derived Fokker-Planck equations will be the same.

1.2.2.1 Derivation from continuous time random walk

Let us consider a simplest example first. Assuming a particle (or walker) moving on a one-dimensional lattice in the way that it moves a distance a to the left or right with equal probability $1/2$ at each time step τ. Now, if the particle is at location x at time $t + \tau$, then it must be at the location $x - a$ or $x + a$ with equal probability $1/2$ at time t. Thus we have

$$p(x, t + \tau) = p(x - a, t)/2 + p(x + a, t)/2. \qquad (1.95)$$

We assume that τ and a are very small so that Eq. (1.95) can be expressed as a Taylor series on (x, t). This procedure yields the partial differential equation

$$\tau \frac{\partial p(x, t)}{\partial t} = \frac{a^2}{2} \frac{\partial^2 p(x, t)}{\partial x^2} + \mathcal{O}(\tau^2) + \mathcal{O}(a^3), \qquad (1.96)$$

where $\mathcal{O}(\tau^2)$ and $\mathcal{O}(a^3)$ are the higher order terms. Letting $a, \tau \to 0$ in the way that

$$D = \lim_{a, \tau \to 0} \frac{a^2}{2\tau} \qquad (1.97)$$

is a positive constant, we obtain the classical heat equation

$$\frac{\partial p(x, t)}{\partial t} = D \frac{\partial^2 p(x, t)}{\partial x^2}, \qquad (1.98)$$

where D is the diffusivity of the particle. If assuming the particle starts its motion from the origin $x = 0$, then the initial condition of Eq. (1.98) is $p(x, 0) = \delta(x)$ and the solution $p(x, t)$ is the Gaussian normal distribution in Eq. (1.76).

The property of moving to the left or right with equal probability $1/2$ is named as unbiased. Sometimes, the particle's motion might be biased as a result of external environment, so that it moves a distance a to the left or right with probabilities l and r, respectively, or stays in the same location with probability $1-l-r$. Similar to the simplest model, we have

$$p(x, t+\tau) = (1-l-r)p(x,t) + rp(x-a,t) + lp(x+a,t). \quad (1.99)$$

The Taylor expansion yields the partial differential equation

$$\tau \frac{\partial p(x,t)}{\partial t} = -a\epsilon \frac{\partial p}{\partial x} + \frac{ka^2}{2} \frac{\partial^2 p(x,t)}{\partial x^2} + \mathcal{O}(\tau^2) + \mathcal{O}(a^3), \quad (1.100)$$

where $\epsilon = r - l$ and $k = r + l$. Letting $a, \tau \to 0$ in the way that

$$v - \lim_{a,\tau \to 0} \frac{a\epsilon}{\tau}, \quad D = k \lim_{a,\tau \to 0} \frac{a^2}{2\tau}, \quad (1.101)$$

are the positive constants, we obtain the drift-diffusion (or advection-diffusion) equation:

$$\frac{\partial p(x,t)}{\partial t} = -v \frac{\partial p}{\partial x} + D \frac{\partial^2 p(x,t)}{\partial x^2}, \quad (1.102)$$

where the first and second term on the right-hand side denote the drift term and diffusion term, respectively. Noting that a^2/τ converges to a positive constant in Eq. (1.101), the probability difference $\epsilon = r - l$ must be proportional to a, and $\epsilon \to 0$ as $a, \tau \to 0$ in Eq. (1.101). Therefore, the probabilities r and l in biased model are not fixed, but vary with the spatial and temporal step sizes such that the limits in Eq. (1.101) exist. If assuming the particle starts its motion from the origin $x = 0$, then the initial condition of Eq. (1.102) is also $p(x, 0) = \delta(x)$ and the solution $p(x, t)$ is

$$p(x,t) = \frac{1}{\sqrt{4\pi Dt}} \exp\left(-\frac{(x-vt)^2}{4Dt}\right), \quad (1.103)$$

being still a Gaussian normal distribution like Eq. (1.76).

The previous model assumes that the particle moves to the left and right with different probability. This assumption yields a global bias of particle's position to left or right. By contrast, there is another model possessing a local bias instead of global bias. In this model, the particle is more likely to move in the same direction to its previous movement direction. This tendency to continue in the same direction is known as persistence.

More precisely, let the particle move along an infinite line at a constant speed v. At each time step τ, the particle either changes the movement direction and moves a distance a in this new direction with probability $\mu\tau$, or moves a distance a in the previous direction with probability $1 - \mu\tau$. In other word, the turning events happen as a Poisson process with rate μ. In this case, the PDF $p(x,t)$ consists of two probability fluxes, the density of right- and left-moving particles at location x and time t denoted by $p_+(x,t)$ and $p_-(x,t)$, respectively. The total PDF satisfies

$$p(x,t) = p_+(x,t) + p_-(x,t). \tag{1.104}$$

Following the description in this model, it holds that

$$
\begin{aligned}
p_+(x,t+\tau) &= (1 - \mu\tau)p_+(x - v\tau, t) + \mu\tau p_-(x + v\tau, t), \\
p_-(x,t+\tau) &= \mu\tau p_+(x - v\tau, t) + (1 - \mu\tau)p_-(x + v\tau, t).
\end{aligned}
\tag{1.105}
$$

Performing the Taylor expansion and taking the limit $\tau \to 0$, we have

$$\frac{\partial p_+}{\partial t} = -v\frac{\partial p_+}{\partial x} + \mu(p_- - p_+), \tag{1.106}$$

and

$$\frac{\partial p_-}{\partial t} = v\frac{\partial p_-}{\partial x} - \mu(p_- - p_+). \tag{1.107}$$

Adding the two equations in Eq. (1.106) and differentiating with respect to t, and subtracting between the two equations in Eq. (1.106) and differentiating with respect to x, respectively, yield

$$\frac{\partial^2(p_+ + p_-)}{\partial t^2} = v\frac{\partial^2(p_- - p_+)}{\partial x \partial t}, \tag{1.108}$$

and

$$\frac{\partial^2(p_- - p_+)}{\partial x \partial t} = v\frac{\partial^2(p_+ + p_-)}{\partial x^2} - 2\mu\frac{\partial(p_- - p_+)}{\partial x}. \tag{1.109}$$

Combining the two equations above, we obtain the equation of $p(x,t)$:

$$\frac{\partial^2 p}{\partial t^2} + 2\mu\frac{\partial p}{\partial t} = v^2\frac{\partial^2 p}{\partial x^2}. \tag{1.110}$$

This equation is also named as telegraph equation, so called because it was originally studied by Lord Kelvin in relation to signals propagating across the transatlantic cable [72]. From the observation of Eq. (1.110), the PDF is indeed unbiased globally. This equation can be solved given both the initial conditions $p(x, 0)$ and $(\partial p/\partial t)(x, 0)$, but the full solution is very complex.

The discussions until here are based on fixed size of time step τ and space step a. In general, at each step, the time cost and the displacement can be random variables, as the waiting time and jump length in CTRW model. Let us assume the power law distributed waiting time and jump length as

$$\psi(t) \simeq A_\alpha(\tau/t)^{1+\alpha}, \quad w(x) \simeq \frac{A^\beta}{\sigma^\beta |x|^{1+\beta}} \qquad (1.111)$$

with $0 < \alpha < 1$ and $0 < \beta < 2$. The coefficients A_α and A^β are chosen properly so that the Laplace transform of waiting time PDF and the Fourier transform of jump length PDF are [97, 187]

$$\hat{\psi}(\lambda) \simeq 1 - \tau^\alpha \lambda^\alpha, \quad \tilde{w}(k) \simeq 1 - \sigma^\beta |k|^\beta, \qquad (1.112)$$

respectively. Substituting them into the Montroll-Weiss equation in Eq. (1.11) gives

$$\tilde{\hat{p}}(k, \lambda) = \frac{\lambda^{\alpha-1}}{\lambda^\alpha + K_\alpha^\beta |k|^\beta} \qquad (1.113)$$

with the generalized diffusion coefficient $K_\alpha^\beta = \sigma^\beta/\tau^\alpha$. Making the deformation on Eq. (1.113) properly yields

$$\lambda \tilde{\hat{p}}(k, \lambda) - 1 = -K_\alpha^\beta \lambda^{1-\alpha} |k|^\beta \tilde{\hat{p}}(k, \lambda). \qquad (1.114)$$

Considering the initial condition $\tilde{p}(k, 0) = 1$, performing the inverse Fourier-Laplace transform on Eq. (1.114), we obtain the Fokker-Planck equation

$$\frac{\partial}{\partial t} p(x, t) = K_\alpha^\beta D_t^{1-\alpha} \nabla_x^\beta p(x, t), \qquad (1.115)$$

where $D_t^{1-\alpha}$ denotes the Riemann-Liouville fractional differential operator, ∇_x^β is the Riesz spatial fractional derivative operator with Lévy exponent $-|k|^\beta$ [31, 188]; and in x space, it is defined as

$$\nabla_x^\beta h(x) = -\frac{_{-\infty}D_x^\beta h(x) + {_x}D_\infty^\beta h(x)}{2 \cos(\beta\pi/2)}, \qquad (1.116)$$

where for $n - 1 < \beta < n$,

$$_{-\infty}D_y^\beta h(x) = \frac{1}{\Gamma(n - \beta)} \frac{d^n}{dx^n} \int_{-\infty}^{x} \frac{h(x')}{(x - x')^{\beta+1-n}} dx', \qquad (1.117)$$

$$_xD_\infty^\beta h(x) = \frac{(-1)^n}{\Gamma(n - \beta)} \frac{d^n}{dx^n} \int_{x}^{\infty} \frac{h(x')}{(x' - x)^{\beta+1-n}} dx'. \qquad (1.118)$$

1.2.2.2 Derivation from Langevin equation

The position of a particle can be directly given through an over-damped Langevin equation

$$\dot{x}(t) = f(x(t), t) + g(x(t), t)\xi(t), \qquad (1.119)$$

where $x(t)$ denotes the particle's coordinate, $f(x, t) = -\partial U(x, t)/\partial x$ a force field, $U(x, t)$ an external potential, $\xi(t)$ a random noise resulting from a fluctuating environment, and $g(x, t)$ a multiplicative noise term implying a heterogeneous environment. Since a random noise is the formal derivative of the corresponding process, we denote $\eta(t) = \int_0^t \xi(t')dt'$ being the noise generating process. For convenience, we assume $\xi(t)$ is the Lévy noise. Therefore, the noise generating process $\eta(t)$ is the Lévy process, which has the independent and stationary increments, as Sec. 1.1.2 says.

Now, let us derive the Fokker-Planck equation from the Langevin equation Eq. (1.119) with the technique of Fourier transform [50]. Considering the displacement of the particle in a time interval $[t, t + \tau]$ with sufficiently small τ, it holds that

$$\delta x(t) \simeq f(x(t), t)\tau + g(x(t), t)\delta\eta(t) \qquad (1.120)$$

in the Itô sense [83, 156], where

$$\delta x(t) = x(t + \tau) - x(t),$$
$$\delta\eta(t) = \int_t^{t+\tau} \xi(t')dt'. \qquad (1.121)$$

The current increment of particle' coordinate $\delta x(t)$ only depends on the current increments $\delta\eta(t)$ of noise generating process $\eta(t)$ and thus it is independent on the previous increments of $\eta(t)$ since the increments of Lévy process is independent on non-overlapping intervals. Because of the stationary increment of the Lévy process, we know that $\delta\eta(t)$ has the same distribution as $\eta(\tau)$ with characteristic function denoted by [12]:

$$\langle e^{-ik\eta(\tau)}\rangle = e^{\tau\phi_0(k)}, \tag{1.122}$$

where the Lévy exponent $\phi_0(k)$ characterizes the jump structure of the Lévy noise $\xi(t)$. In the subsequent part, for a specific Lévy noise, it has the specific form that $\phi_0(k) = -k^2$ for Gaussian white noise and $\phi_0(k) = -|k|^\beta$ $(0 < \beta < 2)$ for non-Gaussian β-stable Lévy noise.

We define the PDF through δ function:

$$p(x,t) = \langle\delta(x - x(t))\rangle, \tag{1.123}$$

and its Fourier transform

$$\tilde{p}(k,t) = \langle e^{-ikx(t)}\rangle. \tag{1.124}$$

Then we express the increment in Fourier domain as

$$\begin{aligned}\delta\tilde{p}(k,t) &= \tilde{p}(k,t+\tau) - \tilde{p}(k,t) \\ &= \langle e^{-ikx(t+\tau)} - e^{-ikx(t)}\rangle.\end{aligned} \tag{1.125}$$

For sufficiently small τ, we have

$$\delta\tilde{p}(k,t) \simeq -ik\tau\langle e^{-ikx(t)}f(x(t),t)\rangle + \langle e^{-ikx(t)}(e^{-ikg(x(t),t)\delta\eta(t)} - 1)\rangle. \tag{1.126}$$

Then we divide the both sides of Eq. (1.126) by τ and taking the limit $\tau \to 0$. The left-hand side becomes

$$\lim_{\tau\to 0}\delta\tilde{p}(k,t) = \frac{\partial\tilde{p}(k,t)}{\partial t}. \tag{1.127}$$

The first term on the right-hand side of Eq. (1.126) is just the Fourier transform of a compound function on $p(x,t)$, i.e.,

$$ik\langle e^{-ikx(t)}f(x(t),t)\rangle = \mathcal{F}_x\left\{\frac{\partial}{\partial x}f(x,t)p(x,t)\right\}. \tag{1.128}$$

While the second term on the right-hand side of Eq. (1.126) yields

$$\lim_{\tau\to 0}\frac{1}{\tau}\langle(e^{-ikg(x(t),t)\delta\eta(t)} - 1)\rangle = \phi_0(kg(x(t),t)), \tag{1.129}$$

by using the characteristic function of the noise increment $\delta\eta(t)$ in Eq. (1.122). Combining Eqs. (1.127), (1.128), and (1.129), we obtain the Fokker-Planck equation in Fourier space:

$$\frac{\partial \tilde{p}(k,t)}{\partial t} = -\mathcal{F}_x\left\{\frac{\partial}{\partial x}f(x,t)p(x,t)\right\} + \mathcal{F}_x\{\phi_0(kg(x,t))p(x,t)\}.$$

$$(1.130)$$

Once the form of $\phi_0(kg(x,t))$ is given for a specific noise, the Fokker-Planck equation in x space is obtained.

If the noise $\xi(t)$ is the Gaussian white noise, then $\phi_0(k) = -k^2$, and the Fokker-Planck equation is

$$\frac{\partial}{\partial t}p(x,t) = -\frac{\partial}{\partial x}f(x,t)p(x,t) + \frac{\partial^2}{\partial x^2}g^2(x,t)p(x,t). \qquad (1.131)$$

If the noise $\xi(t)$ is the Lévy noise, then $\phi_0(k) = -|k|^\beta$, and the Fokker-Planck equation is

$$\frac{\partial}{\partial t}p(x,t) = -\frac{\partial}{\partial x}f(x,t)p(x,t) + \nabla_x^\beta |g(x,t)|^\beta p(x,t), \qquad (1.132)$$

where ∇_x^β is the Riesz spatial fractional derivative operator defined in Eq. (1.116).

If the deterministic time variable in Langevin equation Eq. (1.119) is replaced by a positive non-decreasing one-dimensional Lévy process, called subordinator [12], then the subordinated stochastic process could be described by the following coupled Langevin equation

$$\begin{aligned} \dot{x}(s) &= f(x(s),t(s)) + g(x(s),t(s))\xi(s), \\ \dot{t}(s) &= \theta(s). \end{aligned} \qquad (1.133)$$

Here we adopt the fully skewed α-stable Lévy noise $\theta(s)$ with $0 < \alpha < 1$, which is independent of the arbitrary Lévy noise $\xi(s)$. Then the combined process is defined as $y(t) := x(s(t))$ with the inverse α-stable subordinator $s(t)$, which is the first passage time of the α-stable subordinator $\{t(s), s \geq 0\}$ and is defined [149] as $s(t) = \inf_{s>0}\{s : t(s) > t\}$. Note that the time-dependent force f and multiplicative noise term g should depend on the physical time $t(s)$, rather than the operation time s, to guarantee the physical interpretation [78, 126].

Based on the generalized results of the α-dependent subordinator and inverse subordinator, the PDF $p(y,t)$ of the subordinated process $y(t)$ can be written through the PDF $p_0(y,s)$ of the original process $y(s)$, i.e., [16, 19]

$$p(y,t) = \int_0^\infty p_0(y,s)h(s,t)ds, \qquad (1.134)$$

where $h(s,t)$ is the PDF of inverse subordinator $s(t)$. The original process's PDF $p_0(y,s)$ should satisfy the Fokker-Planck equation

$$\frac{\partial}{\partial t}p_0(y,t) = \mathcal{L}_{\mathrm{FP}}p_0(y,t), \qquad (1.135)$$

where $\mathcal{L}_{\mathrm{FP}}$ is the Fokker-Planck operator. It can be $-\frac{\partial}{\partial x}f(x,t) + \frac{\partial^2}{\partial x^2}g^2(x,t)$ in Eq. (1.131) or $-\frac{\partial}{\partial x}f(x,t) + \nabla_x^\beta |g(x,t)|^\beta$ in Eq. (1.132). This operator acts only on spatial variable, and can be exchanged with a temporal operator. Considering the Eq. (1.68) $h(s,t)$ satisfies, we have

$$\frac{\partial}{\partial t}p(y,t) - \int_0^\infty p_0(y,s)\frac{\partial}{\partial t}h(s,t)ds$$
$$= -\int_0^\infty p_0(y,s)D_t^{1-\alpha}\frac{\partial}{\partial s}h(s,t)ds \qquad (1.136)$$
$$= D_t^{1-\alpha}\int_0^\infty p_0(y,s)\frac{\partial}{\partial s}h(s,t)ds.$$

Then performing the integration by parts on the right-hand side gives

$$\frac{\partial}{\partial t}p(y,t) = D_t^{1-\alpha}\int_0^\infty h(s,t)\frac{\partial}{\partial s}p_0(y,s)ds, \qquad (1.137)$$

where the boundary terms vanish since $h(0,t) = h(\infty,t) = 0$ for any fixed t. Combining the Fokker-Planck equation (1.135), we obtain

$$\frac{\partial}{\partial t}p(y,t) = D_t^{1-\alpha}\int_0^\infty h(s,t)L_{\mathrm{FP}}p_0(y,s)ds. \qquad (1.138)$$

Taking the Fokker-Planck operator L_{FP} out of the integral and using the relation in Eq. (1.134) again, the Fokker-Planck equation for subordinated process $y(t)$ is obtained:

$$\frac{\partial}{\partial t}p(y,t) = D_t^{1-\alpha}L_{\mathrm{FP}}p(y,t). \qquad (1.139)$$

1.3 FUNCTIONAL

We have paid attention to the position of a diffusion process in Sec. 1.2. Only a few PDFs can be explicitly given in real domain. Instead, we derive the Fokker-Planck equation governing the PDFs under the framework of CTRW and Langevin equation. Besides of the position of the diffusion process, people are sometimes more interested in the functional of the sample path of a process, defined as

$$A = \int_0^t U[x(t')]dt', \qquad (1.140)$$

where $U(x)$ is a specified function. The functional A takes different values for different samples of $x(t)$. Therefore, the functional A is also a stochastic process. The different choices of $U(x)$ lead to different meanings of functional A, such as first passage time, occupation time in a given domain and stock price. These specific functionals have been applied to numerous fields, such as probability theory, data analysis, finance, computer science, and disordered systems [127].

It is hard to obtain the explicit expression of PDF of particle's position in real domain, not to mention the PDF of particle's functional. Therefore, people focus to derive the governing equations of the PDF of the functionals, which are named as Feynman-Kac equation. It was firstly derived by Kac in 1949 for normal diffusion [92], influenced by Feynman's thesis about Schrödinger's equation. If we define the PDF of finding the functional taking value A at time t for the particles with initial position x_0 as $G_{x_0}(A, t)$, then its Laplace transform $G_{x_0}(\rho, t)$ $(A \to \rho)$ satisfies the Feynman-Kac equation

$$\frac{\partial G_{x_0}(\rho, t)}{\partial t} = K \frac{\partial^2 G_{x_0}(\rho, t)}{\partial x_0^2} - \rho U(x_0) G_{x_0}(\rho, t). \qquad (1.141)$$

Here, we omit the Laplace notion $\hat{\cdot}$ with respect to the argument of ρ for convenience, since the Feynman-Kac equations are usually presented in the Laplace ρ space. Note that the spatial derivatives in Eq. (1.141) act on the initial position x_0, which makes it named as backward Feynman-Kac equation. The initial condition of Eq. (1.141) is $G_{x_0}(\rho, 0) = 1$, which can be extracted from the definition of functional in Eq. (1.140) that $G_{x_0}(A, 0) = \delta(A)$. The $G_{x_0}(\rho, t)$ in Eq. (1.141) can be uniquely solved for a given initial condition and specific function $U(x)$.

The functionals of the sample path of Brownian particles have been investigated in numerous studies. Especially in recent years, as the accelerating discovery and development of anomalous diffusion processes in many fields, the Feynman-Kac equations governing the PDF of the functional of the anomalous path in complex systems have attracted people's attention. In fact, the classical Feynman-Kac equation in Eq. (1.141) is not valid any more for these anomalous cases. It is urgent to establish a set of theories of new Feynman-Kac equations for these anomalous diffusion processes.

There have been some attempts of deriving the Feynman-Kac equations under the framework of CTRW model [30, 31, 80, 175, 179, 188, 189] Langevin equation [182], and Itô formula [28, 29]. Fortunately, the obtained results are interesting and the ordinary time derivative in Eq. (1.141) is replaced by a fractional substantial derivative [61]. Equation (1.141) can also be recovered within some specific cases. The corresponding derivations will be presented by using three different methods individually.

1.3.1 Derivation from Continuous Time Random Walk

In CTRW framework, the particles move on a one-dimensional infinite lattice. Assume that the particle moves a distance a to the left or right with equal probability $1/2$ at each time step. Waiting times between successive jumps are i.i.d. random variables, which are distributed in the power law form

$$\psi(\tau) \simeq \frac{B_\alpha \tau^{-(1+\alpha)}}{|\Gamma(-\alpha)|} \tag{1.142}$$

with B_α being a constant and $0 < \alpha < 1$. Under this condition, the mean waiting time is infinite, which implies that the particle experiences the trap events frequently.

1.3.1.1 Forward Feynman-Kac equation

Let $G(x, A, t)$ be the joint PDF of finding the particle at position x with the functional value A at time t. By using the technique of transport equation, we introduce the auxiliary function $Q_n(x, A, t)$ to represent the probability of the particle making its nth jump into (x, A) at time t.

Based on the definition of functional in Eq. (1.140), we build the recursion relation for Q_n:

$$Q_{n+1}(x, A, t) = \frac{1}{2} \int_0^t \psi(\tau) Q_n[x - a, A - \tau U(x - a), t - \tau] d\tau$$
$$+ \frac{1}{2} \int_0^t \psi(\tau) Q_n[x + a, A - \tau U(x + a), t - \tau] d\tau. \tag{1.143}$$

Equation (1.143) can be interpreted as follows. For particle making its $(n + 1)$th jump into (x, A) at time t, it must be located at the nearest neighbors $x \pm a$ with equal probability $1/2$ before this jump. The waiting time τ can be any value between 0 and t, and the functional value added within this waiting time is $\tau U(x \pm a)$. For the initial condition $n = 0$ without jumps happened, it holds that $Q_0(x, A, t) = \delta(x)\delta(A)\delta(t)$.

The PDF $G(x, A, t)$ we concern can be represented through the auxiliary function $Q_n(x, A, t)$ as

$$G(x, A, t) = \int_0^t \Phi(\tau) \sum_{n=0}^{\infty} Q_n[x, A - \tau U(x), t - \tau] d\tau, \tag{1.144}$$

where

$$\Phi(t) = 1 - \int_0^t \psi(\tau) d\tau \tag{1.145}$$

is the survival probability. The survival probability $\Phi(t)$ in the integrand of Eq. (1.144) implies that the particle performs its last jump at time $t - \tau$ at position x and stays at position x without jumping before time t. Throughout this waiting time, the functional value added is $\tau U(x)$.

Considering the convolution form in Eqs. (1.143) and (1.144), we will apply the technique of Laplace and Fourier transform to solve them. Assuming that the functional A is positive, we perform the Laplace transform $A \to \rho, t \to \lambda$ and Fourier transform $x \to k$ on both sides of Eq. (1.143) and obtain

$$\hat{\tilde{Q}}_{n+1}(k, \rho, \lambda) = \cos(ka)\hat{\psi}\left[\lambda + \rho U\left(-i\frac{\partial}{\partial k}\right)\right] \hat{\tilde{Q}}_n(k, \rho, \lambda). \tag{1.146}$$

The initial condition $\hat{\tilde{Q}}_0(k, \rho, \lambda) = 1$ can help solve all the $\hat{\tilde{Q}}_n(k, \rho, \lambda)$:

$$\hat{\tilde{Q}}_n(k, \rho, \lambda) = \left[\cos(ka)\hat{\psi}\left[\lambda + \rho U\left(-i\frac{\partial}{\partial k}\right)\right]\right]^n. \tag{1.147}$$

On the other hand, performing the Laplace transform $A \to \rho$ on Eq. (1.144) yields

$$\int_0^\infty Q_n[x, A - U(x)\tau, t - \tau]e^{-\rho A} dA = e^{-\rho U(x)\tau} Q_n(x, \rho, t - \tau).$$

(1.148)

Then performing the Laplace transform $t \to \lambda$ and utilizing the convolution theorem of Laplace transform, we obtain

$$\hat{G}(x, \rho, \lambda) = \sum_{n=0}^\infty \hat{\Phi}(\lambda + \rho U(x))\hat{Q}_n(x, \rho, \lambda),$$

(1.149)

where $\hat{\Phi}(\lambda) = (1 - \hat{\psi}(\lambda))/\lambda$ is the Laplace transform of the survival probability $\Phi(t)$. Further performing the Fourier transform $x \to k$ on Eq. (1.149) and substituting the expression of Q_n in Eq. (1.147) into it, we obtain

$$\tilde{\hat{G}}(k, \rho, \lambda) = \sum_{n=0}^\infty \hat{\Phi}\left[\lambda + \rho U\left(-i\frac{\partial}{\partial k}\right)\right]\tilde{\hat{Q}}_n(k, \rho, \lambda)$$

$$= \frac{1 - \hat{\psi}[\lambda + \rho U(-i\frac{\partial}{\partial k})]}{\lambda + \rho U(-i\frac{\partial}{\partial k})} \cdot \frac{1}{1 - \cos(ku)\hat{\psi}[\lambda + \rho U(-i\frac{\partial}{\partial k})]}.$$

(1.150)

Note that Eq. (1.150) is actually a generalization of the well-known Montroll-Weiss equation [136, 141] to the scenario about functional. Considering the relationship

$$\tilde{\hat{G}}(k, \lambda) = \int_0^\infty \tilde{\hat{G}}(k, A, \lambda)dA = \tilde{\hat{G}}(k, \rho = 0, \lambda),$$

(1.151)

we take $\rho = 0$ in Eq. (1.150), and obtain the marginal PDF of position x

$$\tilde{\hat{G}}(k, \lambda) = \frac{1 - \hat{\psi}(\lambda)}{\lambda} \frac{1}{1 - \cos(ka)\hat{\psi}(\lambda)},$$

(1.152)

which is exactly the Montroll-Weiss equation.

For the concerned power law waiting time distribution, the corresponding $\tilde{\hat{G}}(k, \rho, \lambda)$ can be obtained by using the asymptotic expansion of $\hat{\psi}(\lambda)$ for small λ, i.e.,

$$\hat{\psi}(\lambda) \simeq 1 - B_\alpha \lambda^\alpha.$$

(1.153)

Considering the long time limit of $\lambda, k \to 0$ in Eq. (1.150) and denoting the anomalous diffusion coefficient as

$$K_\alpha = \frac{a^2}{2B_\alpha}, \tag{1.154}$$

we obtain

$$\tilde{G}(k, \rho, \lambda) \sim \left[\lambda + \rho U\left(-i\frac{\partial}{\partial k}\right)\right]^{\alpha-1} \frac{1}{K_\alpha k^2 + \left[\lambda + \rho U\left(-i\frac{\partial}{\partial k}\right)\right]^\alpha}. \tag{1.155}$$

Rearranging the above expression and inverting it back to the real domain result in the forward fractional Feynman-Kac equation

$$\frac{\partial G(x, \rho, t)}{\partial t} = K_\alpha \frac{\partial^2}{\partial x^2} \mathcal{D}_t^{1-\alpha} G(x, \rho, t) - \rho U(x) G(x, \rho, t), \tag{1.156}$$

where $\mathcal{D}_t^{1-\alpha}$ is called the fractional substantial derivative operator [62]. In Laplace space, the fractional substantial derivative operator can be described as $\mathcal{D}_t^{1-\alpha} G(x, \rho, t) \to [\lambda + \rho U(x)]^{1-\alpha} \hat{G}(x, \rho, \lambda)$.

There are some remarks on the forward Feynman-Kac equation (1.156).

1. For normal diffusion with $\alpha = 1$, the fractional equation (1.156) reduces to the classical Feynman-Kac equation

$$\frac{\partial G(x, \rho, t)}{\partial t} = K_1 \frac{\partial^2}{\partial x^2} G(x, \rho, t) - \rho U(x) G(x, \rho, t). \tag{1.157}$$

2. The Feynman-Kac equation (1.156) governs the joint PDF of position x and functional A. By taking $\rho = 0$, it reduces to the fractional Fokker-Planck equation governing the PDF of position x

$$\frac{\partial G(x, t)}{\partial t} = K_\alpha \frac{\partial^2}{\partial x^2} D_t^{1-\alpha} G(x, t), \tag{1.158}$$

where $D_t^{1-\alpha}$ is the Riemann-Liouville fractional derivative operator ($D_t^{1-\alpha} G(x, t) \to \lambda^{1-\alpha} \hat{G}(x, \lambda)$ in Laplace space $t \to \lambda$).

3. If the jump length is also power law distributed as $w(x) \propto |x|^{-1-\beta}$ with $0 < \beta < 2$, the corresponding characteristic

function is $\tilde{w}(k) \simeq 1 - C_\beta |k|^\beta$ and the fractional Feynman-Kac equation is [30]

$$\frac{\partial}{\partial t} G(x, \rho, t) = K_{\alpha,\beta} \nabla_x^\beta \mathcal{D}_t^{1-\alpha} G(x, \rho, t) - \rho U(x) G(x, \rho, t),$$

(1.159)

where $K_{\alpha,\beta} = C_\beta / B_\alpha$, and ∇_x^β is the Riesz spatial fractional derivative operator.

1.3.1.2 Backward Feynman-Kac equation

In contrast to the forward Feynman-Kac equation, the backward Feynman-Kac equation governs the PDF $G_{x_0}(A, t)$ of A at time t given that the particle starts at x_0. The initial position x_0 is a deterministic variable, not a random one here. Since the PDF $G_{x_0}(A, t)$ depends on the initial position x_0 instead of the final position x, and we should pay attention to the first step in the CTRW framework rather than the last step. In detail, the particle jumps to either $x_0 + a$ or $x_0 - a$ with equal probabilities $1/2$ after the waiting time τ in the first step. Then, the remaining steps can be characterized through $G_{x_0 \pm a}(A - \tau U(x_0), t - \tau)$, which is

$$G_{x_0}(A, t) = \frac{1}{2} \int_0^t \psi(\tau) G_{x_0+a}[A - \tau U(x_0), t - \tau] d\tau$$

$$+ \frac{1}{2} \int_0^t \psi(\tau) G_{x_0-a}[A - \tau U(x_0), t - \tau] d\tau \qquad (1.160)$$

$$+ \Phi(t) \delta[A - U(x_0)t].$$

The last term in Eq. (1.160) denotes the particle staying at the initial position x_0 during the measurement time $[0, t]$.

Similar to the procedure of deriving forward Feynman-Kac equation, we perform the Laplace transform $(A \to \rho, t \to \lambda)$ and Fourier transform $x_0 \to k_0$, and obtain

$$\tilde{\hat{G}}_{k_0}(\rho, \lambda) = \hat{\psi}\left[\lambda + \rho U\left(-i\frac{\partial}{\partial k_0}\right)\right] \cos(k_0 a) \tilde{\hat{G}}_{k_0}(\rho, \lambda)$$

$$+ \frac{1 - \hat{\psi}\left[\lambda + \rho U\left(-i\frac{\partial}{\partial k_0}\right)\right]}{\lambda + \rho U\left(-i\frac{\partial}{\partial k_0}\right)} \delta(k_0). \qquad (1.161)$$

We use the asymptotic expansion of $\hat{\psi}(\lambda)$ in Eq. (1.153), and obtain

$$\lambda \tilde{\hat{G}}_{k_0}(\rho, \lambda) - \delta(k_0) = -K_\alpha \left[\lambda + \rho U \left(-i \frac{\partial}{\partial k_0} \right) \right]^{1-\alpha} k_0^2 \tilde{\hat{G}}_{k_0}(\rho, \lambda)$$

$$- \rho U \left(-i \frac{\partial}{\partial k_0} \right) \tilde{\hat{G}}_{k_0}(\rho, \lambda).$$

(1.162)

Inverting the above equation to real domain yields the backward fractional Feynman-Kac equation

$$\frac{\partial}{\partial t} G_{x_0}(\rho, t) = K_\alpha \mathcal{D}_t^{1-\alpha} \frac{\partial^2}{\partial x_0^2} G_{x_0}(\rho, t) - \rho U(x_0) G_{x_0}(\rho, t). \quad (1.163)$$

There are some remarks on the backward Feynman-Kac equation (1.163).

1. Taking $\alpha = 1$, the backward Feynman-Kac equation (1.163) reduces to the classical one (1.141).

2. If the jump length is also power law distributed as $w(x) \propto |x|^{-1-\beta}$ with $0 < \beta < 2$, the corresponding backward fractional Feynman-Kac equation is [30]

$$\frac{\partial}{\partial t} G_{x_0}(\rho, t) = K_{\alpha,\beta} \mathcal{D}_t^{1-\alpha} \nabla_{x_0}^\beta G_{x_0}(\rho, t) - \rho U(x_0) G_{x_0}(\rho, t).$$

(1.164)

3. Comparing the forward fractional Feynman-Kac equation (1.156) and the backward one (1.163), we find the main differences are the variable $x \to x_0$ and the operators $\frac{\partial^2}{\partial x^2} \mathcal{D}_t^{1-\alpha} \to \mathcal{D}_t^{1-\alpha} \frac{\partial^2}{\partial x_0^2}$. Note that the order of the operators cannot be interchanged casually since the fractional substantial derivative operator $\mathcal{D}_t^{1-\alpha}$ depends on the spatial variable.

1.3.2 Derivation from Langevin Equation

Let us derive the Feynman-Kac equation from an overdamped Langevin equation. Compared with the CTRW model, the Langevin equation has the advantage of describing a heterogeneous environment and an external force field by including a multiplicative noise term and a force term

$$\dot{x}(t) = f(x(t), t) + g(x(t), t)\xi(t), \quad (1.165)$$

where $x(t)$ is the particle coordinate, $f(x,t)$ is the force field, $\xi(t)$ is the noise resulting from a fluctuating environment, and $g(x,t)$ is the multiplicative noise term. Similar to the analyses in Sec. 1.2.2, we choose $\xi(t)$ to be the Lévy noise, which is the formal time derivative of the corresponding Lévy process $\eta(t)$. Denote the increment of particle's position and the Lévy process as $\delta\eta(t) = \int_t^{t+\tau} \xi(t')dt'$ and $\delta\eta(t) = \eta(t+\tau) - \eta(t)$ in the time interval $[t, t+\tau]$. Considering the sufficient small τ, we have

$$\delta x(t) \simeq f(x(t),t)\tau + g(x(t),t)\delta\eta(t), \qquad (1.166)$$

where the multiplicative noise is interpreted in the Itô sense interpretation [83, 156]. Let the characteristic function of Lévy process $\eta(t)$ be [12]:

$$\langle e^{-ik\eta(t)} \rangle = e^{t\phi_0(k)}, \qquad (1.167)$$

where the Lévy exponent $\phi_0(k)$ can be $\phi_0(k) = -k^2$ for Gaussian white noise and $\phi_0(k) = -|k|^\beta$ $(0 < \beta < 2)$ for non-Gaussian β-stable Lévy noise.

1.3.2.1 Forward Feynman-Kac equation

Take the previous notation $G(x, A, t)$ to be the joint PDF of position x and functional A at time t. The Fourier transform $x \to k, A \to \rho$ of $G(x, A, t)$ can be denoted by

$$\tilde{G}(k, \rho, t) = \langle e^{-ikx(t)} e^{-i\rho A(t)} \rangle. \qquad (1.168)$$

Similar to the increment $\delta x(t)$ in Eq. (1.166), the increment of functional $A(t)$ is

$$\delta A(t) = A(t+\tau) - A(t) \simeq U(x(t))\tau \qquad (1.169)$$

during the small time interval $[t, t+\tau]$. Then we write the increment of $G(x, A, t)$ in Fourier space as

$$\begin{aligned} \delta\tilde{G}(k, \rho, t) :&= \tilde{G}(k, \rho, t+\tau) - \tilde{G}(k, \rho, t) \\ &= \langle e^{-ikx(t+\tau)-i\rho A(t+\tau)} \rangle - \langle e^{-ikx(t)-i\rho A(t)} \rangle. \end{aligned} \qquad (1.170)$$

Substituting the increments $\delta x(t)$ and $\delta A(t)$ into Eq. (1.170) and taking $\tau \to 0$ give

$$\begin{aligned} \delta\tilde{G}(k, \rho, t) = \ &\langle e^{-ikx(t)-i\rho A(t)}(e^{-ikg(x(t),t)\delta\eta(t)} - 1) \rangle \\ &- ik\tau \langle e^{-ikx(t)-i\rho A(t)} f(x(t),t) \rangle \\ &- i\rho\tau \langle e^{-ikx(t)-i\rho A(t)} U(x(t)) \rangle. \end{aligned} \qquad (1.171)$$

Note that the angle brackets in the first term in Eq. (1.171) denote the average with the joint PDF $G(x, A, t)$ and the PDF of the noise increment $\delta\eta(t)$. Since $\delta\eta(t)$ is independent of particle trajectory $x(t)$, we perform the average on $\delta\eta(t)$ first by utilizing the characteristic function of Lévy process in Eq. (1.167), and obtain

$$\lim_{\tau \to 0} \frac{1}{\tau} \langle (e^{-ikg(x(t),t)\delta\eta(t)} - 1) \rangle = \phi_0(kg(x(t), t)). \qquad (1.172)$$

It can be seen that the second and third terms in Eq. (1.171) are exactly the Fourier transforms of the compound functions with respect to $G(x, A, t)$, i.e.,

$$ik\langle e^{-ikx(t)-i\rho A(t)} f(x(t), t) \rangle = \mathcal{F}_x\mathcal{F}_A \left\{ \frac{\partial}{\partial x} f(x, t)G(x, A, t) \right\}, \qquad (1.173)$$

and

$$i\rho\langle e^{-ikx(t)-i\rho A(t)} U(x(t)) \rangle = i\rho \mathcal{F}_x\mathcal{F}_A \left\{ U(x)G(x, A, t) \right\}. \qquad (1.174)$$

Up to now, the terms on the right-hand side of Eq. (1.171) have been evaluated in Eqs. (1.172), (1.173), and (1.174), respectively. Then dividing Eq. (1.171) by τ and taking the limit $\tau \to 0$, we can obtain the forward Feynman-Kac equation in Fourier space:

$$
\begin{aligned}
\frac{\partial \tilde{G}(k, \rho, t)}{\partial t} &= \mathcal{F}_x\{\phi_0(kg(x, t))G(x, \rho, t)\} \\
&\quad - \mathcal{F}_x\left\{ \frac{\partial}{\partial x} f(x, t)G(x, \rho, t) + i\rho U(x)G(x, \rho, t) \right\}.
\end{aligned}
\qquad (1.175)
$$

There are some remarks on the forward fractional Feynman-Kac equation (1.175).

1. *Fokker-Planck equation.* Let $\rho = 0$ in Eq. (1.175). In this case, $G(x, \rho = 0, t) = \int_0^\infty G(x, A, t)dA$ reduces to the marginal PDF $G(x, t)$ of finding the particle at position x at time t. Correspondingly, the forward Feynman-Kac equation (1.175) reduces to the Fokker-Planck equation (1.130).

2. *Gaussian white noise.* If $\xi(t)$ is the Gaussian white noise in Eq. (1.175), then the forward Feynman-Kac equation is

$$\frac{\partial G(x,\rho,t)}{\partial t} = \left[-\frac{\partial}{\partial x}f(x,t) + \frac{\partial^2}{\partial x^2}g^2(x,t) \right] G(x,\rho,t)$$

$$- i\rho U(x)G(x,\rho,t).$$

(1.176)

3. *Non-Gaussian β-stable noise.* If $\xi(t)$ is the non-Gaussian β-stable noise in Eq. (1.175), then the forward Feynman-Kac equation becomes

$$\frac{\partial G(x,\rho,t)}{\partial t} = \left[-\frac{\partial}{\partial x}f(x,t) + \nabla_x^\beta |g(x,t)|^\beta \right] G(x,\rho,t)$$

$$- i\rho U(x)G(x,\rho,t),$$

(1.177)

where ∇_x^β is the Riesz spatial fractional derivative operator with Lévy exponent $-|k|^\beta$.

4. *A positive functional.* If the functional A is positive at any time t, such as the first passage time and occupation time, the Fourier transform $A \to \rho$ can be replaced by the Laplace transform $G(x,\rho,t) = \int_0^\infty e^{-\rho A}G(x,A,t)dA$. Hence, the forward Feynman-Kac equation corresponding to Eq. (1.175) is obtained by replacing $i\rho$ with ρ.

1.3.2.2 Backward Feynman-Kac equation

Similar to the derivation of the backward Feynman-Kac equation under the framework of CTRW, the PDF of interest is $G_{x_0}(A,t)$ of finding functional value A at time t given that the particle starts at x_0. Since the PDF $G_{x_0}(A,t)$ depends on the initial position x_0 instead of the final position x, we assume the underlying process to be time homogenous, and pay attention to the first step.

In detail, we assume that the stochastic process is

$$\dot{x}(t) = f(x(t)) + g(x(t))\xi(t),$$

(1.178)

where $\xi(t)$ is also a Lévy noise. Compared with the model Eq. (1.165), f and g do not explicitly depend on t. Otherwise, the time-dependent force field (or the multiplicative noise term) induces different displacement for a particle located at the same position but different time. And this case will bring difficulty to letting the functional A only depend on initial position x_0 without using the information of the whole path $x(t)$.

The important thing is that how we build the dependence of the functional A on initial position x_0. Different from the increment δA considered in the forward Feynman-Kac equation, we write the functional in the following way:

$$
\begin{aligned}
A(t+\tau)|_{x_0} &= \int_0^\tau U(x(t'))dt' + \int_\tau^{t+\tau} U(x(t'))dt' \\
&= U(x_0)\tau + A(t)|_{x(\tau)},
\end{aligned}
\tag{1.179}
$$

where $A(t+\tau)|_{x_0}$ denotes the functional A at time $t+\tau$ with the initial position x_0. Letting $t = 0$ in Eq. (1.166), $x(\tau)$ can be approximated as

$$
x(\tau) = x_0 + f(x_0)\tau + g(x_0)\eta(\tau). \tag{1.180}
$$

The expression of $G_{x_0}(A, t)$ in Fourier space is

$$
G_{x_0}(\rho, t) = \langle e^{-i\rho A(t)|_{x_0}} \rangle. \tag{1.181}
$$

Then it holds that

$$
G_{x_0}(\rho, t+\tau) = \langle\langle e^{-i\rho A(t)|_{x(\tau)}} \rangle\rangle e^{-i\rho U(x_0)\tau} \tag{1.182}
$$

based on Eq. (1.179). Since $A(t)|_{x(\tau)}$ denotes the functional A at time t with the initial position $x(\tau)$, it is independent of the event before $x(\tau)$, e.g., $\eta(\tau)$. So the internal angle brackets in Eq. (1.182) denote the average over $A(t)|_{x(\tau)}$ while the external ones the average over $\eta(\tau)$. Then the increment $\delta G_{x_0}(\rho, t)$ can be expressed as

$$
\begin{aligned}
\delta G_{x_0}(\rho, t) :&= G_{x_0}(\rho, t+\tau) - G_{x_0}(\rho, t) \\
&= \langle\langle e^{-i\rho A(t)|_{x(\tau)}} \rangle\rangle e^{-i\rho U(x_0)\tau} - \langle e^{-i\rho A(t)|_{x_0}} \rangle.
\end{aligned}
\tag{1.183}
$$

Taking $\tau \to 0$, omitting the higher order terms of τ, we obtain

$$
\begin{aligned}
\delta G_{x_0}(\rho, t) &= \langle\langle e^{-i\rho A(t)|_{x(\tau)}} \rangle\rangle - \langle e^{-i\rho A(t)|_{x_0}} \rangle \\
&\quad - i\rho U(x_0)\tau \langle e^{-i\rho A(t)|_{x_0}} \rangle,
\end{aligned}
\tag{1.184}
$$

where the last term is equal to $-i\rho U(x_0)\tau G_{x_0}(\rho, t)$. Next, we will deal with the first two terms on the right-hand side of Eq. (1.184) carefully and omit the higher order of τ. Performing Fourier transform $x_0 \to k_0$ on Eq. (1.184), the term $\langle e^{-i\rho A(t)|x_0} \rangle$ yields $\tilde{G}_{k_0}(\rho, t)$. But for $\langle\langle e^{-i\rho A(t)|x(\tau)} \rangle\rangle$, it is not easy to get the form in Fourier space.

Hence, we firstly take $g(x) \equiv 1$, i.e., the noise in this system is additive. Denote $T_\eta = \langle e^{-i\rho A(t)|x(\tau)} \rangle$. Since $g(x) \equiv 1$, Eq. (1.180) becomes

$$x(\tau) = x_0 + f(x_0)\tau + \eta(\tau), \qquad (1.185)$$

where $f(x_0)$ only depends on the initial position x_0. Because of this dependence, when we consider the Fourier transform, $x(\tau)$ is not a simple shift of x_0. Therefore, we firstly write the Fourier transform $(x_0 \to k_0)$ of $\langle T_\eta \rangle$ as

$$\mathcal{F}_{x_0}\{\langle T_\eta \rangle\} = \left\langle \int_{-\infty}^{\infty} e^{-ik_0 x(\tau)} T_\eta e^{ik_0(f(x_0)\tau + \eta(\tau))} dx_0 \right\rangle. \qquad (1.186)$$

Then turning dx_0 into $dx(\tau)$, one arrives at

$$\mathcal{F}_{x_0}\{\langle T_\eta \rangle\} = \left\langle \int_{-\infty}^{\infty} e^{-ik_0 x(\tau)} T_\eta e^{ik_0(f(x_0)\tau + \eta(\tau))} dx(\tau) \right\rangle$$
$$- \left\langle \int_{\infty}^{\infty} e^{-ik_0 x(\tau)} T_\eta e^{ik_0(f(x_0)\tau + \eta(\tau))} \frac{df(x_0)}{dx_0} \tau dx_0 \right\rangle. \qquad (1.187)$$

Since all x_0 and $f(x_0)$ are multiplied by τ in Eq. (1.187), replacing all x_0 by $x(\tau)$ in Eq. (1.187) yields higher-order terms of τ, which can be omitted. Then we utilize the asymptotics $e^{ik_0 f(x_0)\tau} \simeq 1 + ik_0 f(x_0)\tau$. The first term on the right-hand side of Eq. (1.187) reduces to

$$\left\langle \int_{-\infty}^{\infty} e^{-ik_0 x(\tau)} T_\eta e^{ik_0 \eta(\tau)} dx(\tau) \right\rangle$$
$$+ ik_0\tau \left\langle \int_{-\infty}^{\infty} e^{-ik_0 x(\tau)} T_\eta f(x(\tau)) dx(\tau) \right\rangle, \qquad (1.188)$$

where the latter term above equals to

$$\tau \mathcal{F}_{x_0}\left\{ \frac{\partial}{\partial x_0} f(x_0) G_{x_0}(\rho, t) \right\}. \qquad (1.189)$$

On the other hand, omitting the exponential term and replacing x_0 by $x(\tau)$ on the second term on the right-hand side of Eq. (1.187) yield

$$-\tau \Big\langle \int_{-\infty}^{\infty} e^{-ik_0 x(\tau)} T_\eta \frac{df(x(\tau))}{dx(\tau)} dx(\tau) \Big\rangle = -\tau \mathcal{F}_{x_0} \Big\{ \frac{df(x_0)}{dx_0} G_{x_0}(\rho, t) \Big\}.$$
(1.190)

Therefore, by replacing $x(\tau)$ by y, the Fourier transform of $\langle\langle e^{-i\rho A(t)|_{x(\tau)}} \rangle\rangle - \langle e^{-i\rho A(t)|_{x_0}} \rangle$ in Eq. (1.184) reduces to

$$\Big\langle \int_{-\infty}^{\infty} e^{-ik_0 y} T_\eta (e^{ik_0 \eta(\tau)} - 1) dy \Big\rangle + \tau \mathcal{F} \Big\{ f(x_0) \frac{\partial G_{x_0}(\rho, t)}{\partial x_0} \Big\}, \quad (1.191)$$

i.e.,

$$\tau \phi_0(-k_0) \tilde{G}_{k_0}(\rho, t) + \tau \mathcal{F}_{x_0} \Big\{ f(x_0) \frac{\partial G_{x_0}(\rho, t)}{\partial x_0} \Big\} \qquad (1.192)$$

on account of Eq. (1.172). Dividing Eq. (1.184) by τ and taking $\tau \to 0$, we obtain the backward Feynman-Kac equation in Fourier space:

$$\frac{\partial \tilde{G}_{k_0}(\rho, t)}{\partial t} = \phi_0(-k_0) \tilde{G}_{k_0}(\rho, t)$$

$$+ \mathcal{F}_{x_0} \Big\{ f(x_0) \frac{\partial G_{x_0}(\rho, t)}{\partial x_0} - i\rho U(x_0) G_{x_0}(\rho, t) \Big\}.$$
(1.193)

There are some remarks on the backward Feynman-Kac equation (1.193).

1. *Gaussian white noise.* If the noise $\xi(t)$ is Gaussian white noise, then $\phi_0(-k_0) = -k_0^2$ and we get the backward Feynman-Kac equation:

$$\frac{\partial G_{x_0}(\rho, t)}{\partial t} = \frac{\partial^2}{\partial x_0^2} G_{x_0}(\rho, t)$$

$$+ f(x_0) \frac{\partial}{\partial x_0} G_{x_0}(\rho, t) - i\rho U(x_0) G_{x_0}(\rho, t),$$
(1.194)

which is the same as the backward Feynman-Kac equation with $\alpha = 1$ in Ref. [31] obtained under the framework of CTRW.

2. *Non-Gaussian β-stable noise.* If the noise $\xi(t)$ is non-Gaussian β-stable noise, i.e., $\phi_0(-k_0) = -|k_0|^\beta$, then the backward Feynman-Kac equation becomes

$$
\frac{\partial G_{x_0}(\rho, t)}{\partial t} = \nabla_{x_0}^\beta G_{x_0}(\rho, t)
$$
$$
+ f(x_0) \frac{\partial}{\partial x_0} G_{x_0}(\rho, t) - i\rho U(x_0) G_{x_0}(\rho, t),
$$

(1.195)

which is an extension for the backward Feynman-Kac equation derived from CTRW [31], in which the jump length obeys heavy-tailed distribution but without a force field $f(x)$.

3. *Multiplicative noise.* When $g(x)$ is not a constant, we assume $\xi(t)$ to be Gaussian white noise and the backward Feynman-Kac equation is

$$
\frac{\partial G_{x_0}(\rho, t)}{\partial t} = g^2(x_0) \frac{\partial^2}{\partial x_0^2} G_{x_0}(\rho, t)
$$
$$
+ f(x_0) \frac{\partial}{\partial x_0} G_{x_0}(\rho, t) - i\rho U(x_0) G_{x_0}(\rho, t),
$$

(1.196)

which recovers to Eq. (1.194) when $g(x_0) \equiv 1$. See the detailed derivations in [182].

1.3.2.3 Coupled Langevin equation

Similar to Eq. (1.133), we should consider the subordinated stochastic process described by the coupled Langevin equation:

$$
\dot{x}(s) = f(x(s), t(s)) + g(x(s), t(s))\xi(s),
$$
$$
\dot{t}(s) = \theta(s),
$$

(1.197)

where $\theta(s)$ is the fully skewed α-stable Lévy noise with $0 < \alpha < 1$, independent of the arbitrary Lévy noise $\xi(s)$. Then the subordinated stochastic process is defined as $y(t) = x(s(t))$ with the inverse α-stable subordinator $s(t)$, and we can build the Langevin equation of $y(t)$ from Eq. (1.197) as:

$$
\dot{y}(t) = f(y(t), t)\dot{s}(t) + g(y(t), t)\xi(s(t))\dot{s}(t).
$$

(1.198)

In the interpretation of Itô sense, the increment of $y(t)$ can be written as

$$\delta y(t) = f(y(t), t)\delta s(t) + g(y(t), t)\delta \eta(s(t)), \tag{1.199}$$

where $\delta s(t) = s(t + \tau) - s(t)$ and $\delta \eta(s(t)) = \eta(s(t + \tau)) - \eta(s(t))$. Next, similar to Eq. (1.171), we obtain the increment of $G(y, W, t)$ in Fourier space ($y \to k, W \to \rho$):

$$\delta \tilde{G}(k, \rho, t) = \langle e^{-iky(t)-i\rho W(t)}(e^{-ikg(y(t),t)\delta \eta(s(t))} - 1)\rangle$$

$$- ik\langle e^{-iky(t)-i\rho W(t)}f(y(t), t)\delta s(t)\rangle \tag{1.200}$$

$$- i\rho\tau\langle e^{-iky(t)-i\rho W(t)}U(y(t))\rangle,$$

where the first term on the right-hand side can be reduced to

$$\langle e^{-iky(t)-i\rho W(t)}\phi_0(kg(y(t), t))\,\delta s(t)\rangle \tag{1.201}$$

as usual due to the characteristic function of $\delta \eta(t)$ in Eq. (1.167). Then dividing Eq. (1.200) by τ and taking the limit $\tau \to 0$, we obtain

$$\frac{\partial}{\partial t}\tilde{G}(k, \rho, t) = \langle e^{-iky(t)-i\rho W(t)}\phi_0(kg(y(t), t))\dot{s}(t)\rangle$$

$$- ik\langle e^{-iky(t)-i\rho W(t)}f(y(t), t)\dot{s}(t)\rangle \tag{1.202}$$

$$- i\rho\langle e^{-iky(t)-i\rho W(t)}U(y(t))\rangle$$

$$=: Q_1 + Q_2 + Q_3.$$

It is obvious that the inverse Fourier transform ($k \to y$) of Q_3 is $-i\rho U(y)G(y, \rho, t)$. But for Q_1 and Q_2, they look a little bit difficult due to the new term $\dot{s}(t)$ compared with Eq. (1.171). Note that the angle brackets in Q_1 denote the average of two kinds of independent stochastic processes with the joint PDF $G(y(t), W(t), t)$ and Lévy α-stable noise $\theta(t)$ on which $s(t)$ depends. In order to deal with the term Q_1, we first add a technical δ-function $\delta(y - y(t))$ in it and get

$$Q_1 = \int_{-\infty}^{\infty} e^{-iky}\phi_0(kg(y, t))\langle e^{-i\rho W(t)}\delta(y - y(t))\dot{s}(t)\rangle dy. \tag{1.203}$$

Then applying the technique in [29] of rewriting the functional $W(t)$ as a subordinated process:

$$W(t) = V(s(t)), \quad V(s) = \int_0^s U(x(s'))\theta(s')ds'. \tag{1.204}$$

Substituting $y(t) = x(s(t))$ and $W(t) = V(s(t))$ into Q_1 gives the middle term of Q_1 as

$$\langle e^{-i\rho V(s(t))}\delta(y - x(s(t)))\dot{s}(t)\rangle$$
$$= \int_0^\infty \langle e^{-i\rho V(s)}\delta(y - x(s))\delta(t - T(s))\rangle ds. \tag{1.205}$$

Performing Laplace transform $(t \to \lambda)$ on Eq. (1.205), we obtain

$$\hat{Q}_1(\lambda) = \int_{-\infty}^\infty e^{-iky}\phi_0(kg(y,t))\int_0^\infty \langle e^{-i\rho V(s)-\lambda t(s)}\delta(y - x(s))\rangle ds dy. \tag{1.206}$$

On the other hand, $G(y, \rho, t)$ can be rewritten as:

$$G(y, \rho, t) = \langle e^{-i\rho V(s(t))}\delta(y - x(s(t)))\rangle$$
$$= \int_0^\infty \langle e^{-i\rho V(s)}\delta(s - s(t))\delta(y - x(s))\rangle ds. \tag{1.207}$$

So its Laplace transform $(t \to \lambda)$ is

$$\hat{G}(y, \rho, \lambda) = \int_0^\infty \langle e^{-i\rho V(s)-\lambda t(s)}\theta(s)\delta(y - x(s))\rangle ds. \tag{1.208}$$

The characteristic function of the Lévy process $t(s)$ in Eq. (1.197) is

$$\langle e^{-\lambda t(s)}\rangle = e^{-s\lambda^\alpha}. \tag{1.209}$$

Then we use the independence between $\theta(s)$ and $x(s)$ and perform the average with respect to $\theta(s)$ first. Thus the part of the integrand of Eq. (1.208) becomes

$$\langle e^{-i\rho V(s)-\lambda t(s)}\theta(s)\rangle = \left\langle \theta(s)e^{-\int_0^s [\lambda+i\rho U(x(r))]\theta(r)dr}\right\rangle$$
$$= -\frac{1}{\lambda + i\rho U(x(s))}\frac{\partial}{\partial s}\left\langle e^{-\int_0^s [\lambda+i\rho U(x(r))]\theta(r)dr}\right\rangle$$
$$= -\frac{1}{\lambda + i\rho U(x(s))}\frac{\partial}{\partial s}e^{-\int_0^s [\lambda+i\rho U(x(r))]^\alpha dr}$$
$$= [\lambda + i\rho U(x(s))]^{\alpha-1}\left\langle e^{-\int_0^s [\lambda+i\rho U(x(r))]\theta(r)dr}\right\rangle$$
$$= [\lambda + i\rho U(x(s))]^{\alpha-1}\langle e^{-i\rho V(s)-\lambda t(s)}\rangle, \tag{1.210}$$

which yields

$$\hat{G}(y, \rho, \lambda) = [\lambda + i\rho U(y)]^{\alpha-1} \int_0^\infty \langle e^{-i\rho V(s) - \lambda t(s)} \delta(y - x(s)) \rangle ds. \tag{1.211}$$

Comparing Eq. (1.206) with Eq. (1.211), we find that

$$Q_1(\lambda) = \int_{-\infty}^\infty e^{-iky} \phi_0(kg(y, t))[\lambda + i\rho U(y)]^{1-\alpha} G(y, \rho, \lambda) dy. \tag{1.212}$$

Taking the inverse Laplace transform $(\lambda \to t)$, we obtain

$$Q_1 = \int_{-\infty}^\infty e^{-iky} \phi_0(kg(y, t)) \mathcal{D}_t^{1-\alpha} G(y, \rho, t) dy. \tag{1.213}$$

As for Q_2, it can be obtained similarly, i.e.,

$$Q_2 = -ik \int_{-\infty}^\infty e^{-iky} f(y, t) \mathcal{D}_t^{1-\alpha} G(y, \rho, t) dy. \tag{1.214}$$

Finally, substituting Eqs. (1.213) and (1.214) into Eq. (1.202), we obtain the forward Feynman-Kac equation in Fourier space:

$$\frac{\partial \tilde{G}(k, \rho, t)}{\partial t} = \mathcal{F}_y \{\phi_0(kg(y, t)) \mathcal{D}_t^{1-\alpha} G(y, \rho, t)\}$$
$$- \mathcal{F}_y \left\{ \frac{\partial}{\partial y} f(y, t) \mathcal{D}_t^{1-\alpha} G(y, \rho, t) + i\rho U(y) G(y, \rho, t) \right\}. \tag{1.215}$$

1.3.3 Derivation from Itô Formula

An alternative method to derive the forward Feynman-Kac equation (1.215) corresponding the coupled Langevin equation Eq. (1.197) is to use the Itô formula [144] in Ref. [29], which studied the basic mathematic knowledge of Lévy processes, semimartingales and their stochastic calculus. In addition, it also extends the waiting time to be arbitrarily distributed, such as tempered power law distribution. The different waiting time distribution contributes to a different fractional substantial derivative operator.

Let us still use the Langevin equation (1.197), where the underlying process is $y(t) = x(s(t))$ and the functional is denoted as $W(t)$. Then we start from the two dimensional joint process $Z(t) = (y(t), W(t))$. The process $Z(t)$ is a semimartingale with continuous paths, as also $Y(t)$ and $W(t)$ [29]. Thus, the Itô formula is applicable here, which is given explicitly by (for a general smooth function h):

$$h(Z(t)) = h(Z_0) + \int_0^t \frac{\partial}{\partial y} h(Z(\tau))dy(\tau) + \int_0^t \frac{\partial}{\partial W} h(Z(\tau))dW(\tau)$$
$$+ \frac{1}{2} \int_0^t \frac{\partial^2}{\partial y \partial W} h(Z(\tau))d[y, W]_\tau$$
$$+ \frac{1}{2} \int_0^t \frac{\partial^2}{\partial y^2} h(Z(\tau))d[y, y]_\tau$$
$$+ \frac{1}{2} \int_0^t \frac{\partial^2}{\partial W^2} h(Z(\tau))d[W, W]_\tau,$$

$$(1.216)$$

where $[y, W]_t$ is the joint quadratic variation of processes $y(t)$ and $W(t)$. Considering Eq. (1.198), it holds that

$$dy(t) = f(y(\tau), \tau)ds(\tau) + g(y(\tau), \tau)dB(s(\tau)), \qquad (1.217)$$

and

$$y(t) - y_0 = \int_0^t f(y(\tau), \tau)ds(\tau) + \int_0^t g(y(\tau), \tau)dB(s(\tau))$$
$$= \int_0^t f(x(s(\tau)), \tau)ds(\tau) + \int_0^t g(x(s(\tau)), \tau)dB(s(\tau)).$$

$$(1.218)$$

Therefore, the quadratic variation of $y(t)$ is

$$[y, y]_t = [x, x]_{s(t)} = \int_0^{s(t)} g^2(x(\tau), \tau)d\tau$$
$$= \int_0^t g^2(x(s(\tau)), \tau)ds(\tau) = \int_0^t g^2(y(\tau), \tau)\dot{s}(\tau)d\tau.$$

$$(1.219)$$

By contrast, the quadratic variation $[W, W]_t$ and covariation $[y, W]_t$ are both zero since $W(t)$ is a finite variation process. Therefore, recalling the definition of functional implying that $dW(t) = U(y(t))dt$, substituting $dW(t)$, $dy(t)$ in Eq. (1.217) and the quadratic variations into Eq. (1.216), we obtain

$$h(Z(t)) = \int_0^t \frac{\partial}{\partial W} h(Z(\tau))U(y(\tau))d\tau$$

$$+ \int_0^t \frac{\partial}{\partial y} h(Z(\tau))f(y(\tau),\tau)\dot{s}(\tau)d\tau$$

$$+ \int_0^t \frac{\partial}{\partial y} h(Z(\tau))g(y(\tau),\tau)\xi(s(\tau))\dot{s}(\tau)d\tau \qquad (1.220)$$

$$+ \int_0^t \frac{\partial^2}{\partial y^2} h(Z(\tau))g^2(y(\tau),\tau)\dot{s}(\tau)d\tau + h(Z_0).$$

Since $\tilde{G}(k,\rho,t) = e^{-iky(t)-i\rho W(t)}$, we take $h(Z(t)) = e^{-iky(t)-i\rho W(t)}$, and obtain the equation for the double Fourier transform of the joint PDF $\tilde{G}(k,\rho,t)$:

$$h(Z(t)) = -i\rho \int_0^t h(Z(\tau))U(y(\tau))d\tau$$

$$- ik \int_0^t h(Z(\tau))f(y(\tau),\tau)\dot{s}(\tau)d\tau$$

$$- ik \int_0^t h(Z(\tau))g(y(\tau),\tau)\xi(s(\tau))\dot{s}(\tau)d\tau \qquad (1.221)$$

$$- k^2 \int_0^t h(Z(\tau))g^2(y(\tau),\tau)\dot{s}(\tau)d\tau + h(Z_0).$$

By taking the ensemble average on both sides over the realizations of both ξ and θ. The last term on the right-hand side of Eq. (1.221) vanishes due to the independence of the increments of Brownian motion and its independence on the realization of inverse subordinator $s(t)$. Therefore, Eq. (1.221) reduces to

$$\langle h(Z(t)) \rangle = h(Z_0) - i\rho \int_0^t \langle h(Z(\tau)) \rangle U(y(\tau))d\tau$$

$$+ \left\langle \int_0^t h(Z(\tau)) \Big[-ikf(y(\tau),\tau) - k^2 g^2(y(\tau),\tau)\Big]\dot{s}(\tau)d\tau \right\rangle, \qquad (1.222)$$

where in the second integral we find that the inverse Fourier transform of $[-ikf(y(\tau),\tau) - k^2 g^2(y(\tau),\tau)]$ is $\mathcal{L}_{\text{FP}}(y) = -\frac{\partial}{\partial y}f(y) + \frac{\partial^2}{\partial y^2}g^2(y)$ and the inverse Fourier transform of $h(Z(t))$ is $e^{-i\rho W(t)}\delta(y - y(t))$. Performing the inverse Fourier transform on

Eq. (1.222) and taking the derivative with respect to time t, we get

$$\frac{\partial}{\partial t}G(y,\rho,t) = -i\rho U(y)G(y,\rho,t) + \mathcal{L}_{\text{FP}}(y)\frac{\partial}{\partial t} \quad (1.223)$$

$$\left\langle \int_0^t e^{-i\rho W(\tau)}\delta(y-y(\tau))\dot{s}(\tau)d\tau \right\rangle.$$

To obtain the final Feynman-Kac equation, we need to build the relation between the second term on the right-hand side of Eq. (1.223) and $G(y,\rho,t)$. This procedure is similar to the one dealing with the right-hand side of Eq. (1.202). So we can obtain the forward Feynman-Kac equation

$$\frac{\partial}{\partial t}G(y,\rho,t) = -i\rho U(y)G(y,\rho,t) + \mathcal{L}_{\text{FP}}(y)\mathcal{D}_t^{1-\alpha}G(y,\rho,t), \quad (1.224)$$

which is equivalent to Eq. (1.215) with $\phi_0(k) = -k^2$.

1.4 MEAN SQUARED DISPLACEMENT

In the recent decades, it is widely recognized that anomalous diffusion is a very general phenomenon in the natural world, which is characterized by the nonlinear evolution of MSD with respect to time, i.e., $\langle x^2(t)\rangle \propto t^\gamma$ with $\gamma \neq 1$. Since the $\langle x^2(t)\rangle$ is the result of averaging over a large number of different particle trajectories, it is also called ensemble averaged mean squared displacement, which is abbreviated "EAMSD" in this section. In physics experiments, however, it is sometimes inconvenient to measure a large number of particle trajectories. Therefore, people often pay attention to another very important statistic—time averaged mean squared displacement (TAMSD), which can be obtained through only one trajectory of the particle, and is defined as [135]

$$\overline{\delta^2(\Delta)} = \frac{1}{T-\Delta}\int_0^{T-\Delta}[x(t+\Delta) - x(t)]^2 dt. \quad (1.225)$$

In general, the lag time Δ is required to be much shorter than the total measurement time T to obtain good statistics. Nowadays, single particle tracking has become a powerful tool to study transport processes in cellular membranes [159] and probe the microrheology of the cytoplasm [186, 194]. It can be used to evaluate the time averaged observables in cells through video microscopy of fluorescently labeled molecules. For an ergodic system, the time average is equal to the ensemble average at long time limit, i.e.,

$$\langle x^2(\Delta)\rangle = \lim_{T\to\infty} \overline{\delta^2(\Delta)}. \tag{1.226}$$

But for anomalous diffusion, especially for the molecules diffusing in living cells, the time average often becomes a random variable and irreproducible for the individual trajectories. Then we call the stochastic process non-ergodic or ergodicity breaking.

From the Montroll-Weiss equation (1.11), the EAMSD can be directly obtained with the formula [98]

$$\mathcal{L}[\langle x^2(t)\rangle] = -\frac{\partial^2}{\partial k^2}\tilde{\hat{p}}(k,\lambda)\bigg|_{k=0}. \tag{1.227}$$

In contrast to the simple way of calculating EAMSD from Montroll-Weiss equation, TAMSD mainly depends on the correlation function of the underlying process. Instead, the generalized Green-Kubo formula becomes an effective method to evaluate TAMSD of anomalous processes.

1.4.1 Green-Kubo Formula

The Green-Kubo formula is a central result of nonequilibrium statistical mechanics first discussed by Taylor [172]. It expresses the spatial diffusion coefficient D as a time integral of the stationary velocity correlation function (VCF) [73, 104]:

$$D = \int_0^\infty \langle v(t+\tau)v(t)\rangle d\tau. \tag{1.228}$$

For a Brownian particle of mass m with Stokes' friction, the VCF is exponential

$$\langle v(t+\tau)v(t)\rangle = \frac{k_B T}{m}e^{-\gamma\tau}, \tag{1.229}$$

and we can obtain the Einstein relation $D = k_B T/(m\gamma)$. The Green-Kubo formula offers a simple way of relating the diffusive properties of a system to its velocity dynamics. However, it is only valid when the VCF tends to a constant and its integral is finite. For many anomalous diffusion processes, by contrast, the VCFs decay slowly or are nonstationary, and the Green-Kubo formula needs to be generalized.

Consider a system whose VCF has the following asymptotic scaling form when both t and τ are large compared to the system's microscopic time scales

$$C_v(t+\tau,t) \simeq Ct^{\nu-2}\phi\left(\frac{\tau}{t}\right), \qquad (1.230)$$

where $C > 0$ is a constant and $\nu > 1$. Since $C_v(t+\tau,t)$ depends on time t explicitly, it is an aging correlation function, in contrast to a stationary correlation function that only depends on lag time τ. The asymptotic EAMSD for this type of correlation function is given by

$$\langle x^2(t)\rangle \simeq 2C \int_0^t \int_0^{t_2} t_1^{\nu-2}\phi\left(\frac{t_2-t_1}{t_1}\right) dt_1 dt_2, \qquad (1.231)$$

where we have used the symmetry of VCF. Introducing the variable $s = (t_2 - t_1)/t_1$, we obtain

$$\langle x^2(t)\rangle \simeq 2C \int_0^t \int_0^\infty (s+1)^{-\nu}\phi(s)ds t_2^{\nu-1} dt_2. \qquad (1.232)$$

Thus, we obtain

$$\langle x^2(t)\rangle \simeq 2D_\nu t^\nu, \qquad (1.233)$$

where the diffusivity is

$$D_\nu = \frac{C}{\nu} \int_0^\infty (s+1)^{-\nu}\phi(s)ds. \qquad (1.234)$$

While the usual Green-Kubo relation in Eq. (1.228) holds for normal diffusion, the scaling of Green-Kubo relation in Eq. (1.230) is applicable for $\nu > 1$, which corresponds to superdiffusion. Here, the diffusivity is also expressed as an integral over a function of a single variable. Determining the diffusive behavior of a system from its correlation function thus amounts to determining the exponent ν and the scaling function $\phi(s)$.

As for the TAMSD, we should consider the relative EAMSD with respect to some long aging time t_0:

$$
\begin{aligned}
\langle \Delta x^2(t)\rangle_{t_0} &= \langle (x(t_0+t) - x(t_0))^2\rangle \\
&= \int_{t_0}^{t_0+t} \int_{t_0}^{t_0+t} C_v(t_2,t_1)dt_1 dt_2 \\
&\simeq 2C \int_0^t \int_0^{t_2} (t_1+t_0)^{\nu-2}\phi\left(\frac{t_2-t_1}{t_1+t_0}\right).
\end{aligned}
\qquad (1.235)
$$

Introducing the variables $s = (t_2 - t_1)/(t_1 + t_0)$ and $z = t_2/t$, we can write it as

$$\langle \Delta x^2(t) \rangle_{t_0} \simeq 2 D_v^{t/t_0} t^\nu \qquad (1.236)$$

with

$$D_v^{t/t_0} = \mathcal{C} \int_0^1 z^{\nu-1} \left(1 + \frac{1}{zt/t_0} \right)^{\nu-1} \int_0^{zt/t_0} (s+1)^{-\nu} \phi(s) ds dz. \qquad (1.237)$$

In the limit $t \ll t_0$, the diffusivity D_v^{t/t_0} is governed by the small-s expansion of the scaling function $\phi(s)$:

$$D_v^{t/t_0} = \mathcal{C} \left(\frac{t_0}{t} \right)^{\nu-1} \int_0^1 \int_0^{zt/t_0} \phi(s) ds dz. \qquad (1.238)$$

Generally, if the second moment of the velocity either is asymptotically constant or increases with time, $\langle v^2(t) \rangle \simeq a \mathcal{C} t^\beta$, with some positive constant a and $0 \le \beta < \nu - 1$, continuity demands $\phi(s) \simeq c_l s^{-\delta_l}$ with $\delta_l = 2 - \nu + \beta$ for small s, and thus,

$$D_v^{t/t_0} \simeq \frac{c_l \mathcal{C}}{(\nu - \beta - 1)(\nu - \beta)} \left(\frac{t_0}{t} \right)^\beta. \qquad (1.239)$$

Therefore, the relative EAMSD is

$$\langle \Delta x^2(t) \rangle_{t_0} \simeq 2 \frac{c_l \mathcal{C}}{(\nu - \beta - 1)(\nu - \beta)} t_0^\beta t^{\nu-\beta}. \qquad (1.240)$$

Substituting it into the definition of TAMSD in Eq. (1.225), we obtain the ensemble averaged TAMSD as

$$\langle \overline{\delta^2(\Delta)} \rangle \simeq \frac{2 c_l \mathcal{C}}{(\beta + 1)(\nu - \beta - 1)(\nu - \beta)} t^\beta \Delta^{\nu-\beta}. \qquad (1.241)$$

1.4.2 Ergodic and Aging Behavior

Now we take a special process, intermittent search process, as an example, to calculate its EAMSD and TAMSD by use of the Green-Kubo formula, and to investigate its ergodic and aging properties. Since the observation time might not be the beginning of a process in experiments, aging behavior should be paid some attention and it may display interesting phenomena in anomalous diffusion processes [8, 165]. The intermittent search process switches between two phases, local Brownian search phase and ballistic relocation phase (Lévy walk). The searcher displays a slow reactive motion in the first phase, during which the target can be detected. The latter fast phase aims at relocating into unvisited regions to reduce oversampling, during which the searcher is unable to detect the target.

The diffusion behavior of Lévy walk depends on the exponent α of the power-law distributed running time. It displays ballistic diffusion for $\alpha < 1$ and sub-ballistic superdiffusion for $1 < \alpha < 2$. While for Brownian motion, the particle undergoes normal diffusion with diffusivity D. We assume the particle switches between Lévy walk phase and Brownian phase, denoted as states "$+$" and "$-$", respectively. The velocities of the two-state process are, respectively, $v_+(t)$ for Lévy walk and $v_-(t)$ for Brownian motion. The PDF of $v_+(t)$ is $\delta(|v| - v_0)/2$, while $v_-(t) = \sqrt{2D}\xi(t)$ with $\xi(t)$ being a Gaussian white noise satisfying $\langle \xi(t) \rangle = 0$ and $\langle \xi(t_1)\xi(t_2) \rangle = \delta(t_1 - t_2)$. By taking the diffusivity $D = 0$, the Brownian phase becomes a trap event and we immediately obtain the corresponding process describing the transport of the neuronal messenger ribonucleoproteins.

Let the sojourn times t in the two states "$\pm$" be random variables obeying power-law distribution.

$$\psi_\pm(t) \simeq \frac{a_\pm}{|\Gamma(-\alpha_+)| t^{1+\alpha_\pm}} \tag{1.242}$$

for large t, where $a_\pm$ are scale factors and $\Gamma(\cdot)$ is the Gamma function. We assume that the exponents $\alpha_\pm \in (0, 2)$ in two states and the sojourn times in two states are mutually independent. As usual, we apply the approach of Laplace transform $\hat{\psi}_\pm(\lambda) := \int_0^\infty e^{-\lambda t}\psi_\pm(t)dt$ and obtain the asymptotic behavior of the sojourn time distribution for small λ [139]:

$$\hat{\psi}_\pm(\lambda) \simeq 1 - a_\pm\lambda^{\alpha_\pm} \qquad \text{for} \quad \alpha_\pm \in (0, 1), \tag{1.243}$$

$$\hat{\psi}_\pm(\lambda) \simeq 1 - \mu_\pm\lambda + a_\pm\lambda^{\alpha_\pm} \quad \text{for} \quad \alpha_\pm \in (1, 2), \tag{1.244}$$

where $\mu_\pm$ is the mean sojourn time in state "$\pm$", being finite when $\alpha_\pm \in (1, 2)$. The survival probability that the sojourn time in state "$\pm$" exceeds t is defined as $\Psi_\pm(t) = \int_t^\infty dt'\psi_\pm(t')$ with Laplace transform $\hat{\Psi}_\pm(\lambda) = [1 - \hat{\psi}_\pm(\lambda)]/\lambda$. Note that the dynamic behaviors of standard Lévy walk are significantly different for exponent less or larger than 1 [197]. We will fully discuss the EAMSD and

TAMSD of the two-state process for different sets of $\alpha_\pm$ in the following. Although the mean sojourn time is finite (i.e., $\alpha_\pm > 1$) in most cases, such as the intermittent search process, there are still some circumstances presenting scale free dynamics with $\alpha_\pm < 1$, for example, the RNA transport in neuronal systems. Here we make uniform discussions with $\alpha_\pm \in (0, 2)$ for comprehensive understanding of the two-state process.

Suppose that the particles are initialized at the origin. The propagator $p(x, t)$ represents the PDF of finding the particle at position x at time t. For the two-state process, it is natural to concern which state the particles are located in at time t. Here we denote the joint PDF of finding the particle at position x and state "$\pm$" at time t as $p_\pm(x, t)$, which is associated with the propagator by the relation $p(x, t) = p_+(x, t) + p_-(x, t)$. The subscript "$\pm$" will imply an identical meaning for other quantities.

The integral equations for $p_\pm(x, t)$ can be similarly obtained as the master equations for CTRWs. Besides the sojourn time distribution $\psi_\pm(t)$ and survival probability $\Psi_\pm(t)$, we introduce the notation $G_\pm(x, t)$ to represent the conditional probability that a particle makes a displacement x during sojourn time t at one step in state "$\pm$". Their expressions are given by

$$G_+(x, t) = \delta(|x| - v_0 t)/2, \tag{1.245}$$

$$G_-(x, t) = \frac{1}{\sqrt{4\pi D t}} \exp\left(-\frac{x^2}{4Dt}\right), \tag{1.246}$$

since the state "$+$" represents Lévy walk and state "$-$" denotes Brownian motion, respectively. Then the transport equation governing flux of particles $\gamma_\pm(x, t)$, which defines how many particles leave the position x and change from state "$\mp$" to state "$\pm$" per unit time, can be obtained. This equation connects the flux at the current point to the flux from the neighboring points in the past:

$$\gamma_\pm(x, t) = \int_0^t dt' \int_{-\infty}^\infty dx' \psi_\mp(t') G_\mp(x', t') \gamma_\mp(x - x', t - t') \tag{1.247}$$
$$+ p_\mp^0 \psi_\mp(t) G_\mp(x, t),$$

where we assume that the initial condition is $p_\pm(x, t = 0) = p_\pm^0 \delta(x)$ and the constant $p_\pm^0$ is the initial fraction of two states. The first term on the right-hand side shows that the particles could arrive at position x and state "$\pm$" from another point $x - x'$ after a displacement x' in state "$\mp$". While the second term denotes that the particles initialized in state "$\mp$" turn to state "$\pm$" after a complete step in state "$\mp$". The current density $p_\pm(x, t)$ of particles is connected to the neighboring flux $\gamma_\pm(x, t)$

$$
\begin{aligned}
p_\pm(x, t) = & \int_0^t dt' \int_{-\infty}^{\infty} dx' \Psi_\pm(t') G_\pm(x', t') \gamma_\pm(x - x', t - t') \\
& + p_\pm^0 \Psi_\pm(t) G_\pm(x, t).
\end{aligned}
\tag{1.248}
$$

The last term on the right-hand side accounts for the particles initialized in state "+" staying in this state until the observation time t. The first term sums over all possible neighboring flux $\gamma_\pm(x - x', t - t')$ that could arrive at the point x after displacement x' in state "$\pm$". By means of the techniques of Laplace and Fourier transform, $\tilde{p}_\pm(k, \lambda)$ can be obtained (see the details in Ref. [183]). Besides, the occupation fraction of two states $p_\pm(t)$ as the marginal density of finding the particles in state "$\pm$" at time t can be obtained by taking $k = 0$ in $\tilde{p}_+(k, \lambda)$. The expression of $p_+(t)$ in Laplace space $(t \rightarrow \lambda)$ is

$$
\hat{p}_\pm(\lambda) = \frac{p_\pm^0 + p_\mp^0 \hat{\psi}_\mp(\lambda)}{1 - \hat{\psi}_+(\lambda)\hat{\psi}_-(\lambda)} \cdot \frac{1 - \hat{\psi}_\pm(\lambda)}{\lambda},
\tag{1.249}
$$

the normalization of which can be confirmed by verifying $\hat{p}_+(\lambda) + \hat{p}_-(\lambda) = 1/\lambda$.

Now we analyze the VCF $\langle v(t_1)v(t_2)\rangle$ to calculate the EAMSD and TAMSD of the aging process $x_{t_a}(t)$. The age t_a means that this process has evolved for a time period t_a before we start to observe it, and t is the measurement time. Since the model we consider contains two states: Lévy walk and Brownian motion, represented by symbols "+" and "−", respectively. The VCF could be written as a sum of four possible cases in terms of different states:

$$
\begin{aligned}
\langle v(t_1)v(t_2)\rangle = & \langle v_+(t_1)v_+(t_2)\rangle + \langle v_-(t_1)v_-(t_2)\rangle \\
& + \langle v_+(t_1)v_-(t_2)\rangle + \langle v_-(t_1)v_+(t_2)\rangle.
\end{aligned}
\tag{1.250}
$$

The first term on the right-hand side represents the case that the velocity process $v(t)$ at two time points t_1 and t_2 are both in Lévy walk phase; other terms stand for similar parts of the correlation function. All terms on the right-hand side can be explicitly obtained. For the first term, the velocity is correlated only when there is no renewal happening between t_1 and t_2. Thus, we have

$$\langle v_+(t_1)v_+(t_2)\rangle = v_0^2 p_+(t_1)p_{+,0}(t_1, t_2), \qquad (1.251)$$

where $p_+(t)$ has been given in Eq. (1.249) and $p_{+,0}(t_1, t_2)$ is the PDF that no renewal happens between times t_1 and t_2 in state "+". Similarly, the second term on the right-hand side of Eq. (1.250) is

$$\langle v_-(t_1)v_-(t_2)\rangle = 2D\delta(t_1 - t_2)p_-(t_1)p_{-,0}(t_1, t_2), \qquad (1.252)$$

where $p_{-,0}(t_1, t_2) = 1$ for $t_1 = t_2$, since there must be no renewals within a zero lag time. The two states at times t_1 and t_2 are different in the last two terms on the right-hand side of Eq. (1.250). Therefore, the velocity at t_1 and t_2 are independent. Considering the velocity is unbiased at any time, the last two terms are void.

It can be noted that the PDFs $p_\pm(t)$ and $p_{+,0}(t_1, t_2)$ should be calculated firstly to obtain the VCF in Eq. (1.250). The former one has been given in Eq. (1.249), while the latter one can be derived by using the method in Ref. [71]. The double Laplace transform $(t_1 \to \lambda_1, t_2 \to \lambda_2)$ of $p_{+,0}(t_1, t_2)$ is

$$\hat{p}_{+,0}(\lambda_1, \lambda_2) = \frac{1 + \hat{\psi}_+(\lambda_1 + \lambda_2) - \hat{\psi}_+(\lambda_1) - \hat{\psi}_+(\lambda_2)}{\lambda_1 \lambda_2 (1 - \hat{\psi}_+(\lambda_1 + \lambda_2))}. \qquad (1.253)$$

Based on Eqs. (1.249) and (1.253), the VCF $\langle v(t)v(t + \tau)\rangle$ in Eq. (1.250) can be obtained for different sojourn time distributions $\psi_\pm(t)$. Noticing the asymptotic forms of $p_\pm(t)$ and $p_{+,0}(t, t+\tau)$ for large t, the VCF can be rewritten in the scaling form as

$$\langle v(t)v(t + \tau)\rangle = \langle v_+(t)v_+(t + \tau)\rangle + \langle v_-(t)v_-(t + \tau)\rangle$$
$$\simeq C_1 t^{\nu_1 - 2} \rho\left(\frac{\tau}{t}\right) + C_2 t^{\nu_2 - 1}\delta(\tau), \qquad (1.254)$$

where the parameters ν_1, ν_2 and the scaling function $\rho(\cdot)$ are determined by $p_\pm(t)$ and $p_{+,0}(t, t + \tau)$. The scaling form in Eq. (1.254) helps to show different scaling behaviors of $\langle v(t)v(t+\tau)\rangle$ for different sojourn time distributions $\psi_\pm(t)$, and brings availability to the Green-Kubo formula [44, 137]. For weak aging $t_a \ll t$ and strong aging $t_a \gg t$ cases, it behaves as

$$\langle x_{t_a}^2(t)\rangle \simeq \begin{cases} K_1 t^{\nu_1} + K_2 t^{\nu_2}, & t_a \ll t, \\ K_3 t_a^\beta t^{\nu_1 - \beta} + C_2 t_a^{\nu_2 - 1} t, & t_a \gg t, \end{cases} \qquad (1.255)$$

where the coefficients $K_1 = 2C_1/\nu_1 \int_0^\infty dt(t+1)^{-\nu_1}\rho(t)$, $K_2 = C_2/\nu_2$ and $K_3 = 2c_1 C_1[(\nu_1 - \beta - 1)(\nu_1 - \beta)]^{-1}$. Here c_1 depends on the asymptotic form of scaling function $\rho(z) \simeq c_1 z^{-\delta_1}$ for small z, and β is the exponent of the variance of velocity in the Lévy walk phase for large t [44], i.e.,

$$\langle v_+^2(t) \rangle = v_0^2 p_+(t) \propto t^\beta. \tag{1.256}$$

When constructing single particle tracking experiments, the aging process $x_{t_a}(t)$ is evaluated in terms of its TAMSD, which is defined as

$$\overline{\delta_{t_a}^2(\Delta)} = \frac{1}{T - \Delta} \int_{t_a}^{t_a+T-\Delta} [x(t + \Delta) - x(t)]^2 dt$$

with Δ denoting the lag time and T the total measurement time [135]. If the TAMSD $\overline{\delta_{t_a}^2(\Delta)}$ is equal to the corresponding EAMSD $\langle x_{t_a}^2(\Delta) \rangle$ for sufficiently large T, then this process is ergodic. Similar to the procedure of calculating EAMSD, we obtain the ensemble-averaged TAMSD as:

$$\langle \overline{\delta_{t_a}^2(\Delta)} \rangle \simeq \begin{cases} \frac{K_3}{1+\beta} T^\beta \Delta^{\nu_1-\beta} + K_2 T^{\nu_2-1}\Delta, & t_a \ll T, \\ K_3 t_a^\beta \Delta^{\nu_1-\beta} + C_2 t_a^{\nu_2-1}\Delta, & t_a \gg T. \end{cases} \tag{1.257}$$

There are at least four findings being worth to report from the observations of the generic results of EAMSDs in Eq. (1.255) and ensemble-averaged TAMSDs in Eq. (1.257):

1. All the four mentioned formulae consist of two parts (one from Lévy walk phase and another one from Brownian phase). The exponents of evolution time t or lag time Δ in these two parts might be different from the ones of the corresponding individual Lévy walk and Brownian motion. This is because the PDF $p_\pm(t)$ in Eq. (1.249) plays a weighted role on Lévy walk and Brownian motion. Besides, the sums of exponents of the time variables (including t, t_a, T, Δ) in individual two parts are ν_1 and ν_2, respectively, whatever it is EAMSD or ensemble-averaged TAMSD, and weak or strong aging cases.

2. The exponents of time variables in weak and strong aging cases are closely related for ensemble-averaged TAMSD in Eq. (1.257). While keeping the exponents of Δ invariant and replacing measurement time T by age t_a, the result for strong

aging case is obtained from the one of weak aging case. In other words, the ensemble-averaged TAMSD for weak aging case only depends on T and Δ, while in the same way it counts on t_a and Δ for strong aging cases.

3. The EAMSD and ensemble-averaged TAMSD in weak aging case do not depend on the age t_a, the results of which are identical to the non-aging case $t_a = 0$. In contrast, they explicitly depend on t_a for strong aging case, which implies that the exponents β and $\nu_2 - 1$ of t_a must be zero if the equilibrium initial ensemble (i.e., $t_a \to \infty$ discussed in last section) of this system exists (see specific case 2 in Table 1.1). And in this case, the ensemble-averaged TAMSD will be the same for weak and strong aging cases, and only depends on Δ.

4. Comparing the strong aging EAMSD and the ensemble-averaged TAMSD in Eq. (1.257), it can be noted that

$$\langle x_{t_a}^2(\Delta) \rangle = \langle \overline{\delta_{t_a}^2(\Delta)} \rangle \quad \text{for } t_a \gg T, \tag{1.258}$$

which shows that the aging seemingly makes the weak ergodicity breaking system to be ergodic. It is clear that for any α_- Brownian motion is ergodic in its own phase. However, for TAMSD in Lévy walk phase, there are some differences between $\alpha_+ < 1$ and $1 < \alpha_+ < 2$. For $1 < \alpha_+ < 2$, the mean sojourn time in Lévy walk phase is finite, individual trajectories become self-averaging at sufficiently long (infinite) times, such that there will be no difference between $\overline{\delta_{t_a}^2(\Delta)}$ obtained from different trajectories and ensemble-averaged quantity $\langle \overline{\delta_{t_a}^2(\Delta)} \rangle$ [63, 70]. While for $\alpha_+ < 1$, the characteristic time scale is infinite, then the individual TAMSD $\overline{\delta_{t_a}^2(\Delta)}$ is irreproducible and inequivalent with the EAMSD.

Since both α_+ and α_- go through the range $(0, 2)$, it can be divided into six cases as shown in Table 1.1. However, they can be organized into three categories to deepen understandings of the two-state process by considering the properties of its ingredients, Lévy walk and Brownian motion. It is well-known that the standard Lévy walk performs ballistic diffusion when the exponent of the distribution of running times $\alpha < 1$ and sub-ballistic superdiffusion when $1 < \alpha < 2$, which is faster than the normal diffusion

Table 1.1 Values of several major parameters of EAMSD and TAMSD in Eqs. (1.255) and (1.257) for six cases with different $\alpha_\pm$

Specific cases	ν_1	ν_2	β
1. $\alpha_+ = \alpha_- < 1$	2	1	0
2. $1 < \alpha_\pm < 2$	$3 - \alpha_+$	1	0
3. $\alpha_+ < \alpha_- < 1$	2	$\alpha_+ - \alpha_- + 1$	0
4. $\alpha_+ < 1 < \alpha_- < 2$	2	α_+	0
5. $\alpha_- < \alpha_+ < 1$	$\alpha_- - \alpha_+ + 2$	1	$\alpha_- - \alpha_+$
6. $\alpha_- < 1 < \alpha_+ < 2$	$\alpha_- - \alpha_+ + 2$	1	$\alpha_- - 1$

of Brownian motion. Based on this understanding, the Brownian phase undoubtedly suppresses the diffusion behavior of Lévy walk. This effect may be durable or transient, which is completely determined by the fraction of two states $p_\pm(t)$, or more essentially, the magnitude of the exponents $\alpha_\pm$. The explicit expressions of the MSDs of the six cases can be found in Ref. [183]. Here, we divide them into three categories and analyze their statistical properties:

1. α_+ and α_- are comparable, including the first two cases in Table 1.1. A stationary of the fractions of two states $p_\pm(t)$ can be achieved for long times, that is, $p_\pm(t)$ tends to a constant not equal to 0 or 1. Then the EAMSD and ensemble averaged TAMSD are the combination of the fraction of analogues of individual Lévy walk and Brownian motion whether it is weak aging or strong aging.

2. α_+ is smaller, including the middle two cases in Table 1.1. Due to $\alpha_+ < \alpha_-$, it holds that $p_+(t) \to 1$ as $t \to \infty$, and the Lévy walk phase in state "+" tends to occupy the whole time. Then the results are naturally similar to an individual Lévy walk, except for the asymptotic form of small Δ resulting from Brownian phase.

3. α_- is smaller, including the last two cases in Table 1.1. Due to $\alpha_+ > \alpha_-$, it holds that $p_-(t) \to 1$ as $t \to \infty$, and the Lévy walk phase in state "+" gradually withdraws from the two states in a power-law way. This power-law way suppresses the diffusion of Lévy walk phase and gives the opportunity to Brownian motion to be the leading term when $\alpha_+ - \alpha_- > 1$.

In conclusion, the fraction $p_\pm(t)$ in a two-state process plays a crucial role. It contributes a power term of Δ to weak aging EAMSD, a power term of T to weak aging TAMSD, and a power term of t_a to strong aging EAMSD and TAMSD.

The model Lévy walk interrupted by rest has attracted a lot of attentions in physics and biology. This process has been observed through the experiments on the flow in a rotating annulus as probed by tracer particles [166] and the frictionless motion of a particle in an "egg-crate" potential in a Hamiltonian system [99] many years ago. Recently, it is also observed in the transport of the neuronal messenger ribonucleoproteins delivered to their target synapses [167]. The EAMSD and TAMSD for this model can be obtained by taking the diffusivity D in Brownian phase to be zero. It has been pointed that all the results above consist two parts corresponding to Lévy walk and Brownian motion, respectively. Taking $D = 0$ just eliminates the latter part and brings no effect on the former part of Lévy walk phase. For Lévy walk interrupted by rest, the asymptotic behavior of small Δ in TAMSD disappears and subdiffusion behavior might exist if $\alpha_+ - \alpha_- > 1$.

1.5 MISCELLANEOUS ONES

The most common statistical quantity is the MSD, which has been explicitly studied in Sec. 1.4. Besides, there are many other kinds of quantities concerned in a wide range of problems across different fields ranging from probability theory, finance, data analysis, disordered system, to computer science [127]. Taking the Lévy flight as an example, it has an infinite MSD, but a finite fractional moments $\langle |x(t)|^q \rangle$ with q smaller than the corresponding Lévy exponent. Therefore, fractional moments are complementary to the second moment. Especially, for some anomalous diffusion processes, their fractional moments have different scaling forms for different range of q.

In addition, many practical quantities can be defined through the functional defined in Eq. (1.140) by specifying the function $U(x)$. In probability theory, the occupation time denotes the time spent by a particle in some specific domain within a given time window. It is an important quantity of interest [111]. In this case, $U(x)$ is just a δ-function or an indicator function $\chi(x)$. In finance, the stock price $S(t)$ is usually modelled by the geometric Brownian motion [196], i.e., the exponential of a Brownian motion $S(t) = e^{-\beta x(t)}$ with a positive constant β. This means that $U(x)$ is an exponential function here. The Asian option price [69] depends on the time average of the stock's history price, i.e., $A = \int_0^t e^{-\beta x(t')} dt'$.

Besides of the occupation time and the stock price, there are still some functionals of interest can be obtained with different $U(x)$, such as, the maximal displacement of a diffusing particle within a given time window, the first passage time denoting the time when a particle starting at x_0 first hits another given point, the hitting probability denoting the probability of a particle starting at $0 < x_0 < L$ to hit L before hitting 0, the survival probability of a particle in a medium with an absorbing interval, the area under the random walk curve, etc [31]. In the following, we will provide the detailed analyses on fractional moments and first passage properties with some underlying processes.

1.5.1 Fractional Moments

The common examples of anomalous diffusion processes are the subdiffusive CTRW with divergent first moment of waiting time [27,77] and the Lévy flight with divergent second moment of jump length [164,176]. The common feature of the two typical anomalous diffusion processes is their single mode of the motions. However, a particle moving in a complex or even seemingly simple structures might present simultaneous modes [33, 148], such as the tracing particle under the effect of a flow acting in the phase space of chaotic Hamiltonian systems [68,99]. Such a system is not easy to be analyzed since it exhibits at least two modes of the motion. The common tool to analyze it is the spectrum of exponents $q\nu(q)$ [33] by measuring the absolute q-th moment ($q > 0$) of the displacement of the particles

$$\langle |x(t)|^q \rangle \propto t^{q\nu(q)}. \tag{1.259}$$

For the motions with single mode, $\nu(q)$ is a constant being independent of q, such as $\nu(q) \equiv 1/2$ for Brownian motion. Otherwise, one can find a nonlinear function $\nu(q)$ of q for the motions with multiple modes; this phenomenon is named as strong anomalous diffusion [33].

There have been vast systems exhibiting strong anomalous diffusion, such as the nonlinear dynamical systems [13, 14, 33, 40, 158], the annealed or quenched Lévy walk [10, 21, 26, 71, 160], sand pile models [32, 142, 191], the active transport of polymeric particles in living cells [65], and the spreading of cold atoms in optical lattices [43, 94, 95]. The mechanisms of the strong anomalous diffusion for Lévy walk are studied in detail in Refs. [152–154], where the PDF consists of two kinds of distributions—Lévy distribution in the central part and infinite density in the tail part. The infinite density is non-normalizable, the concept of which was thoroughly investigated as mathematical issues [1, 173], and has been successfully applied to physics; for Lévy walk, it aims at characterizing the ballistic scaling ($x \sim t$), which is complementary to the Lévy scaling in the central part of Lévy walk. In contrast, the propagators of subdiffusive CTRW and Lévy flight only have a single mode, being the stretched Gaussian asymptotics and Lévy distribution [136], respectively. Compared with Lévy flight with divergent MSD, the infinite density characterizes the strong correlation between long jump and long rest in Lévy walk, resulting in a finite MSD. In addition, the infinite density can be used to study the rare fluctuations of occupation time statistics in ergodic CTRW [161] and renewal theory [180]. It is also discussed together with infinite-ergodic theory, for example, the Brownian motion in a logarithmic potential [7] and the Langevin system with multiplicative noise [110, 185].

Here, we turn our attention to the strong anomalous diffusion behavior of the two-state process studied in Sec. 1.4.2, which alternates between standard Lévy walk and Brownian motion. With the similar procedure in Sec. 1.4.2, we can obtain the explicit expression of the PDF in two states as

$$\tilde{p}_{\pm}(k, \lambda) \simeq \frac{\tilde{\tilde{\Phi}}_{\pm}(k, \lambda)}{1 - \tilde{\tilde{\phi}}_{+}(k, \lambda)\tilde{\tilde{\phi}}_{-}(k, \lambda)}. \tag{1.260}$$

Based on Eq. (1.260), the explicit expression of propagator $\tilde{p}(k, \lambda) = \tilde{p}_{+}(k, \lambda) + \tilde{p}_{-}(k, \lambda)$ can be obtained. It can be found that the main ingredients of $\tilde{p}(k, \lambda)$ in Eq. (1.260) are the sojourn time distributions $\psi_{\pm}(t)$. So the further analyses on $\tilde{p}(k, \lambda)$ will be developed for the specific $\psi_{\pm}(t)$ with fixed $\alpha_{\pm}$. Comparing with the thoroughly investigated strong anomalous diffusion behavior of standard Lévy walk, a larger exponent α_{-} in our two-state process makes no difference on the diffusion behavior. Therefore, we only focus on the case of $0 < \alpha_{-} < \alpha_{+} < 1$ here. The similar case of $0 < \alpha_{-} < 1 < \alpha_{+}$ is discussed in Ref. [184], and another four cases will present the same results as pure Lévy walk.

In the case of $0 < \alpha_- < \alpha_+ < 1$, it holds that $\hat{\psi}_\pm(\lambda) \simeq 1 - a_\pm \lambda^{\alpha_\pm}$. Substituting it into Eq. (1.260), we obtain the asymptotic form as $(\lambda, k \to 0)$:

$$\tilde{\hat{p}}(k,\lambda) \simeq \frac{a_-(\lambda + Dk^2)^{\alpha_- - 1} + \frac{a_+}{2}[(\lambda + ikv_0)^{\alpha_+ - 1} + (\lambda - ikv_0)^{\alpha_+ - 1}]}{a_-(\lambda + Dk^2)^{\alpha_-} + \frac{a_+}{2}[(\lambda + ikv_0)^{\alpha_+} + (\lambda - ikv_0)^{\alpha_+}]}.$$

$$(1.261)$$

The normalization of the propagator $p(x,t)$ can be verified by taking $k = 0$ in Eq. (1.261), which yields $\hat{p}(0,\lambda) \simeq 1/\lambda$. The direct inverse Fourier-Laplace transform of Eq. (1.261) is infeasible, which implies extra efforts are needed to deal with Eq. (1.261). Actually, the information contained in the asymptotic form of $\tilde{\hat{p}}(k,\lambda)$ could be extracted through some appropriate scaling analyses. By carefully looking at the denominator in Eq. (1.261), three kinds of scaling ($\lambda \sim k$, $\lambda \sim |k|^2$, and $\lambda \sim |k|^{\alpha_+/\alpha_-}$) can be observed. We will first consider the scaling $\lambda \sim k$, since it characterizes the ballistic scaling of Lévy walk due to its unidirectional flight at each sojourn in this phase.

The outmost distance that particles can arrive at is $\pm v_0 t$ in Lévy walk phase, linear with time, which truncates the PDF $p(x,t)$ at $\pm v_0 t$. Although the particle in Brownian phase might go farther than the distance $\pm v_0 t$, the corresponding distribution decays exponentially when $x \gg \sqrt{t}$ as the propagator $G_-(x,t)$ shows in Eq. (1.245). So we omit the contributions of Brownian phase in the scaling $\lambda \sim k$. The explicit tail information is described by the ballistic scaling $\lambda \sim k$ in Eq. (1.261), corresponding to $x \sim t$ in space-time domain. In contrast to $\lambda \sim k$, another two kinds of scalings aim at characterizing the central part of the graph of the PDF $p(x,t)$.

1.5.1.1 Infinite density of rare fluctuations

To consider the scaling $\lambda \sim k$, we let $\lambda, k \to 0$ and λ/k be fixed. Then Eq. (1.261) can be rewritten as

$$\tilde{\hat{p}}(k,\lambda) \simeq \frac{1}{\lambda} \frac{1 + \frac{a_+}{2a_-}\lambda^{\alpha_+ - \alpha_-}[(1 + \frac{ikv_0}{\lambda})^{\alpha_+ - 1} + (1 - \frac{ikv_0}{\lambda})^{\alpha_+ - 1}]}{1 + \frac{a_+}{2a_-}\lambda^{\alpha_+ - \alpha_-}[(1 + \frac{ikv_0}{\lambda})^{\alpha_+} + (1 - \frac{ikv_0}{\lambda})^{\alpha_+}]}$$

$$(1.262)$$

after neglecting the higher order term k^2. The two terms in Eq. (1.262) containing $\lambda^{\alpha_+ - \alpha_-}$ tend to zero since $\alpha_+ > \alpha_-$. Thus, we further have the asymptotic form as

$$\tilde{\hat{p}}(k,\lambda) \simeq \frac{1}{\lambda} + \frac{a_+}{2a_-}\lambda^{\alpha_+ - \alpha_- - 1}$$

$$\times \left[\left(1 + \frac{ikv_0}{\lambda}\right)^{\alpha_+ - 1} + \left(1 - \frac{ikv_0}{\lambda}\right)^{\alpha_+ - 1} \right.$$

$$\left. - \left(1 + \frac{ikv_0}{\lambda}\right)^{\alpha_+} - \left(1 - \frac{ikv_0}{\lambda}\right)^{\alpha_+} \right] \qquad (1.263)$$

$$= \frac{1}{\lambda} + \frac{a_+}{2a_-}[\tilde{\hat{R}}_\alpha(k,\lambda) + \tilde{\hat{R}}_\alpha(-k,\lambda)],$$

where for convenience we use the notation:

$$\tilde{\hat{R}}_\alpha(k,\lambda) := \lambda^{\alpha_+ - \alpha_- - 1}\left[\left(1 + \frac{ikv_0}{\lambda}\right)^{\alpha_+ - 1} - \left(1 + \frac{ikv_0}{\lambda}\right)^{\alpha_+}\right]$$

$$= \frac{-ikv_0}{\lambda^{\alpha_- + 1}}(\lambda + ikv_0)^{\alpha_+ - 1}. \qquad (1.264)$$

The leading term in Eq. (1.263) is $1/\lambda$, the inverse Fourier-Laplace transform of which is $\delta(x)$. It contributes to a normalized PDF in this scaling while the latter two terms provide the information on the tail of the PDF $p(x,t)$. We consider the ballistic scaling $\lambda \sim k$ here, which compresses all the information in the central part into the origin and thus yields the normalized term $\delta(x)$. Since we are focusing on the information in the tail $|x| > 0$, we omit the term $\delta(x)$ and pay attention to $\tilde{\hat{R}}_\alpha(\pm k,\lambda)$ in Eq. (1.263). With some technical calculations in Ref. [184], the inverse of $\tilde{\hat{R}}_\alpha(\pm k,\lambda)$ can be obtained and we have

$$p(x,t) \simeq \frac{a_+}{2a_- v_0} \frac{t^{\alpha_- - \alpha_+ - 1}}{\Gamma(\alpha_- + 1)\Gamma(1 - \alpha_+)}\mathcal{I}\left(\frac{|x|}{v_0 t}\right), \qquad (1.265)$$

where

$$\mathcal{I}(z) = \mathbf{1}_{(0 < z \leq 1)} z^{-\alpha_+ - 1}(1 - z)^{\alpha_- - 1}[\alpha_+ + (\alpha_- - \alpha_+)z]. \qquad (1.266)$$

There is a truncation at $z = 1$ in the expression of $\mathcal{I}(z)$, which implies $|x| \leq v_0 t$, consistent to the previous analysis that the particle will not go beyond the distance $\pm v_0 t$. On the other hand, regarding $z = x/v_0 t$ as a new variable, the integral of the auxiliary function $\mathcal{I}(z)$ diverges due to its singularity at the origin $z = 0$, which gives it a name—infinite density. Therefore, the infinite density $\mathcal{I}(z)$ is not a real physical PDF. Despite of this, it reveals the long time asymptotic behavior of the propagator $p(x, t)$ through the relationship in Eq. (1.265). When calculating moments, we multiply $|x|^q$ on both sides of Eq. (1.265) and integrate with respect to x. Then we obtain

$$\int_{-\infty}^{\infty} |x|^q p(x, t) dx \propto t^{\alpha_- - \alpha_+ + q} \int_0^1 z^q \mathcal{I}(z) dz. \qquad (1.267)$$

For $q < \alpha_+$, the integral on the right-hand side of Eq. (1.267) diverges. However, the infinite density $\mathcal{I}(z)$ is valid for high order moments with $q > \alpha_+$, which cures the singularity at $z = 0$. Therefore, the main functions of the infinite density $\mathcal{I}(z)$ is to characterize the tail information of PDF $p(x, t)$ and to calculate the high order moments.

We also observe another interesting phenomenon—an accumulation at $z = 1$ due to $\alpha_- < 1$ in Eq. (1.266). This accumulation even exists for $D = 0$ (i.e., the Lévy walk interrupted by rest [99, 166, 167]). Therefore, this accumulation is not contributed by the particles in Brownian phase, which vanishes when $\alpha_- = 1$. While for $\alpha_- < 1$ and big t, it can be balanced by the prefactor $t^{\alpha_- - \alpha_+ - 1}$ in Eq. (1.265). Actually, this phenomenon implies that the PDF of pure Lévy walk is dropped down by the long sojourn time in Brownian phase except for the end point at $z = 1$. The end point of the infinite density is not affected since it results from the particles running in its first step for the whole time. Once the particle renews and turns into the second step in Brownian phase with longer sojourn time, it is less likely for the particle to go back to the Lévy walk phase again.

1.5.1.2 Dual scaling regimes in the central part

After obtaining the tail information of $p(x, t)$ by introducing an infinite density $\mathcal{I}(z)$, we turn our attention to the central part of $p(x, t)$ where the scaling relation $\lambda \ll k$ is valid. This scaling helps to simplify the Eq. (1.261) into

$$\tilde{p}(k, \lambda) \simeq \frac{a_-(\lambda + Dk^2)^{\alpha_- - 1} + \frac{a_+}{2}[(ikv_0)^{\alpha_+ - 1} + (-ikv_0)^{\alpha_+ - 1}]}{a_-(\lambda + Dk^2)^{\alpha_-} + \frac{a_+}{2}[(ikv_0)^{\alpha_+} + (-ikv_0)^{\alpha_+}]}.$$
$$(1.268)$$

It can be found that two different scalings coexist in Eq. (1.268), i.e., $\lambda \sim |k|^{\alpha_+/\alpha_-}$ and $\lambda \sim |k|^2$. This phenomenon is different from the standard Lévy walk, where only Lévy scaling is observed at the central part [152]. Now the Gaussian shape with scaling $\lambda \sim |k|^2$ cannot be omitted due to the longer sojourn time in Brownian phase.

Therefore, it is necessary to consider the magnitude relation between α_+ and $2\alpha_-$ for further analyses. If $\alpha_+ < 2\alpha_-$, the Lévy scaling $\lambda \sim |k|^{\alpha_+/\alpha_-}$ dominates the PDF $p(x,t)$. In this case, after omitting Dk^2 and the second term in numerator of Eq. (1.268) due to $|k|^{\alpha_+-1} \ll \lambda^{\alpha_--1}$, we obtain

$$
\begin{aligned}
\tilde{\hat{p}}(k,\lambda) &\simeq \frac{a_-\lambda^{\alpha_--1}}{a_-\lambda^{\alpha_-} + a_+\cos(\pi\alpha_+/2)v_0^{\alpha_+}|k|^{\alpha_+}} \\
&= \frac{\lambda^{\alpha_--1}}{\lambda^{\alpha_-} + K_\alpha|k|^{\alpha_+}},
\end{aligned}
\tag{1.269}
$$

where the generalized diffusion coefficient $K_\alpha = a_+\cos(\pi\alpha_+/2)v_0^{\alpha_+}/a_-$. When $\alpha_- = 1$, the inverse of $\tilde{\hat{p}}(k,\lambda)$ is a normalized symmetric Lévy stable PDF, which recovers the central part of the PDF of standard Lévy walk. For $\alpha_- < 1$, the PDF is like a Lévy flight coupled with an inverse subordinator. The displacement in Brownian phase can be neglected and thus it acts like a trap event with power law exponent $\alpha_- < 1$. The running time in Lévy walk phase is far less than the one in Brownian phase and thus can be neglected so that the displacement in this phase acts like a jump obeying power law distribution with exponent α_+. The corresponding Langevin system can be found in Ref. [60]. The PDF $p(x,t)$ in Eq. (1.269) is a stretched Lévy distribution, the closed form of which can be expressed by Fox H-function:

$$
p(x,t) \simeq \frac{1}{\sqrt{\pi}|x|} H_{2,3}^{2,1}\left[\frac{|x|^{\alpha_+}}{2^{\alpha_+}K_\alpha t^{\alpha_-}} \,\middle|\, \begin{array}{l} (1,1),(1,\alpha_-) \\ (\frac{1}{2},\frac{\alpha_+}{2}),(1,1),(1,\frac{\alpha_+}{2}) \end{array} \right].
$$

Based on the asymptotic form of Fox H-function [130], there is

$$
p(x,t) \simeq \frac{\tilde{K}_\alpha t^{\alpha_-}}{|x|^{1+\alpha_+}}
\tag{1.270}
$$

for large $|x|$, where the coefficient

$$
\tilde{K}_\alpha = \frac{\Gamma(1+\alpha_+)\sin(\pi\alpha_+/2)K_\alpha}{\Gamma(1+\alpha_-)\pi}.
$$

On the contrary, if $\alpha_+ > 2\alpha_-$, the dominant part of $\tilde{\hat{p}}(k, \lambda)$ is in the scaling $\lambda \sim |k|^2$. Similarly, the second terms in numerator and denominator of Eq. (1.268) are both higher order than the corresponding first terms. We neglect them and obtain

$$\tilde{\hat{p}}(k, \lambda) \simeq \frac{1}{\lambda + Dk^2}, \qquad (1.271)$$

displaying the classical behavior of Brownian motion. The inverse Fourier-Laplace transform of Eq. (1.271) yields the Gaussian shape

$$p(x, t) \simeq \frac{1}{\sqrt{4\pi Dt}} \exp\left(-\frac{x^2}{4Dt}\right) \qquad (1.272)$$

in the central part of $p(x, t)$.

Compared with the infinite density $\mathcal{I}(z)$, the different asymptotic forms of $p(x, t)$ on the central part within different scaling regimes are both normalized, since taking $k = 0$ both yield $\hat{p}(0, \lambda) \simeq 1/\lambda$ in Eqs. (1.269) and (1.271). However, the high order (bigger than α_+) moments will diverge if we use the asymptotic forms of $p(x, t)$ in the central part, since the large-x behavior in Eq. (1.270) is heavy-tailed with exponent $1 + \alpha_+$. On the contrary, the high order moments with Gaussian PDF in Eq. (1.272) exponentially decay and can be neglected compared with the infinite density $\mathcal{I}(z)$ on the tail part.

1.5.1.3 *Complementarity among different scaling regimes*

For both cases of $0 < \alpha_- < \alpha_+ < 1$ and $0 < \alpha_- < 1 < \alpha_+$, the PDFs $p(x, t)$ are studied in different scaling regimes. The tail part $(x \sim t)$ can be well approximated by the infinity density $\mathcal{I}(z)$ as

$$p(x, t) \simeq t^{\alpha_- - \alpha_+ - 1} g_{\text{tail}}\left(\frac{x}{t}\right), \qquad (1.273)$$

where the scaling function

$$g_{\text{tail}}(z) = \frac{a_+}{2a_- v_0 \Gamma(\alpha_- + 1)|\Gamma(1 - \alpha_+)|} \mathcal{I}\left(\frac{|z|}{v_0}\right). \qquad (1.274)$$

On the other hand, the central part of $p(x, t)$ is well approximated by two densities as

$$p(x, t) \simeq \begin{cases} t^{-\alpha_-/\alpha_+} g_{\text{cen1}}\left(xt^{-\alpha_-/\alpha_+}\right) & \alpha_+ < 2\alpha_-, \\ t^{-1/2} g_{\text{cen2}}\left(xt^{-1/2}\right) & \alpha_+ > 2\alpha_-, \end{cases} \qquad (1.275)$$

where the scaling functions

$$g_{\text{cen1}}(z) = \frac{1}{\sqrt{\pi}|z|} H_{2,3}^{2,1} \left[\frac{|z|^{\alpha_+}}{2^{\alpha_+} K_\alpha} \middle| \begin{array}{l} (1,1), (1,\alpha_-) \\ (\frac{1}{2}, \frac{\alpha_+}{2}), (1,1), (1,\frac{\alpha_+}{2}) \end{array} \right] \quad (1.276)$$

and

$$g_{\text{cen2}}(z) = \frac{1}{\sqrt{4\pi D}} \exp\left(-\frac{z^2}{4D}\right). \quad (1.277)$$

The central part of $p(x,t)$ is with the scaling $x \sim t^\beta$, where

$$\beta = \max(\alpha_-/\alpha_+, 1/2) < 1. \quad (1.278)$$

Therefore, the intermediate region between central part t^β and tail part t is very large as $t \to \infty$. For convenience, we simplify Eq. (1.275) as

$$p(x,t) \simeq t^{-\beta} g_{\text{cen}}\left(\frac{x}{t^\beta}\right), \quad (1.279)$$

where $g_{\text{cen}} = g_{\text{cen1}}$ when $\alpha_+ < 2\alpha_-$ and $g_{\text{cen}} = g_{\text{cen2}}$ when $\alpha_+ > 2\alpha_-$.

A natural expectation on the analyses in different scales is that the different distributions should be consistent in the intermediate region. The intermediate region is described by $x \to 0$ for tail part and $x \to \infty$ for central part. By taking the corresponding limits in Eqs. (1.273) and (1.275), respectively, we obtain the same asymptotic form as

$$\begin{aligned} p(x,t) &\simeq t^{\alpha_- - \alpha_+ - 1} g_{\text{tail}}\left(\frac{x}{t}\right) \\ &\simeq t^{-\alpha_-/\alpha_+} g_{\text{cen1}}\left(\frac{x}{t^{\alpha_-/\alpha_+}}\right) \\ &\simeq \frac{c_0 t^{\alpha_-}}{|x|^{1+\alpha_+}} \end{aligned} \quad (1.280)$$

for $t^\beta \ll x \ll t$, where the coefficient

$$c_0 = \frac{a_+ \alpha_+ v_0^{\alpha_+}}{2a_- \Gamma(1+\alpha_-)|\Gamma(1-\alpha_+)|}.$$

The different scaling regimes are complementary here, and they together depict the whole graph of PDF $p(x,t)$. Note that we only use the first density in the central part which decays at a power law rate in Eq. (1.275), since another one decays exponentially and can be omitted. Apart from the consistency of two distributions in the intermediate region, the previous discussions of different dominant roles in Eq. (1.275) make sense when calculating low order moments.

1.5.1.4 Ensemble averages

Now we pay attention to the absolute moments of all orders for the displacement. Since the different scaling regimes approximate the different parts of $p(x,t)$, they together yield the entire information on the long time asymptotics, and thus determine the moments of displacement. We introduce an auxiliary function $c(t)$ which satisfies

$$t^\beta \ll c(t) \ll t \tag{1.281}$$

to divide the central part and tail part. Then we can split the integral into two parts, where different scaling regimes well approximate $p(x,t)$, that is,

$$
\begin{aligned}
\langle |x(t)|^q \rangle &= \int_{|x| \le c(t)} |x|^q p(x,t) dx + \int_{|x| > c(t)} |x|^q p(x,t) dx \\
&= \int_{|x| \le c(t)} |x|^q t^{-\beta} g_{\mathrm{cen}}\left(\frac{x}{t^\beta}\right) dx \\
&\quad + \int_{|x| > c(t)} |x|^q t^{\alpha_- - \alpha_+ - 1} g_{\mathrm{tail}}\left(\frac{x}{t}\right) dx \\
&= t^{\beta q} \int_{|z| \le c(t)/t^\beta} |z|^q g_{\mathrm{cen}}(z) dz \\
&\quad + t^{\alpha_- - \alpha_+ + q} \int_{|z| > c(t)/t} |z|^q g_{\mathrm{tail}}(z) dz.
\end{aligned}
\tag{1.282}
$$

Therefore, the central and tail parts have different contributions to the absolute q-th moments, which are $t^{\beta q}$ and $t^{\alpha_- - \alpha_+ + q}$, respectively, the critical value of which is

$$q_c = \frac{\alpha_+ - \alpha_-}{1 - \beta}, \tag{1.283}$$

implying the piecewise linear behavior of the spectrum of exponents $q\nu(q)$ in Eq. (1.259). When $q < q_c$, the former one plays a leading role, otherwise the latter one dominates. The two integrals in Eq. (1.282) are both finite by choosing appropriate $c(t)$ for different order q. For example, for low order moments with $q < q_c$, choosing $c(t) = c_1 t$ with $c_1 \ll 1$, the two integrals become

$$\int_{-\infty}^{\infty} |z|^q g_{\mathrm{cen}}(z) dz, \quad \int_{|z| > c_1} |z|^q g_{\mathrm{tail}}(z) dz < \infty \qquad (1.284)$$

as $t \to \infty$. The singular point $z = 0$ of the latter integral is excluded by a small distance c_1. While for high order moments with $q > q_c$, we choose $c(t) = c_2 t^\beta$ with $1 \ll c_2$. Then the two integrals are

$$\int_{|z| \le c_2} |z|^q g_{\mathrm{cen}}(z) dz, \quad \int_{-\infty}^{\infty} |z|^q g_{\mathrm{tail}}(z) dz < \infty. \qquad (1.285)$$

The high order moments for the infinity density $g_{\mathrm{tail}}(z)$ will not diverge.

Considering $\beta = \max(\alpha_-/\alpha_+, 1/2)$, the absolute q-th moments are given in two different cases. If $\alpha_+ < 2\alpha_-$,

$$\langle |x(t)|^q \rangle \simeq \begin{cases} M_1^{\prec} t^{q\alpha_-/\alpha_+}, & q < \alpha_+, \\ M^{\succ} t^{q+\alpha_- - \alpha_+}, & q > \alpha_+. \end{cases} \qquad (1.286)$$

If $\alpha_+ \ge 2\alpha_-$,

$$\langle |x(t)|^q \rangle \simeq \begin{cases} M_2^{\prec} t^{q/2}, & q < 2(\alpha_+ - \alpha_-), \\ M^{\succ} t^{q+\alpha_- - \alpha_+}, & q > 2(\alpha_+ - \alpha_-). \end{cases} \qquad (1.287)$$

The results in Eqs. (1.286) and (1.287) imply that this system exhibits strong anomalous diffusion behavior with a bilinear spectrum of exponents. The diffusion coefficients $M_1^{\prec}$, $M_2^{\prec}$, and $M^{\succ}$ can be obtained from the derivations in Eq. (1.282) as

$$
\begin{aligned}
M_1^{\prec} &= \int_{-\infty}^{\infty} |z|^q g_{\mathrm{cen}1}(z) dz \\
&= \frac{(K_\alpha)^{q/\alpha_+} \Gamma(1 - q/\alpha_+) \Gamma(1 + q/\alpha_+)}{\cos(q\pi/2) \Gamma(1 - q) \Gamma(1 + q\alpha_-/\alpha_+)}, \\
M_2^{\prec} &= \int_{-\infty}^{\infty} |z|^q g_{\mathrm{cen}2}(z) dz \\
&= \frac{(4D)^{q/2} \Gamma(\frac{q+1}{2})}{\sqrt{\pi}}, \\
M^{\succ} &= \int_{-\infty}^{\infty} |z|^q g_{\mathrm{tail}}(z) dz \\
&= \frac{a_+ q \Gamma(q - \alpha_+)}{2a_- \Gamma(q - \alpha_+ + \alpha_- + 1) |\Gamma(1 - \alpha_+)|}.
\end{aligned}
\qquad (1.288)
$$

1.5.2 First Passage Time and First Hitting Time

The first passage problem consists in the determination of boundary crossing dynamics of a searcher. In the classical one-dimensional setting, we are interested in the event when a searcher crosses the origin for the first time after initially being released at position $x_0 > 0$. For Lévy flight, the first passage time corresponds to the moment in time when the searcher first hits a coordinate on the negative semi-axis. For Lévy walk, the first passage time is defined as the time needed for a particle to reach the origin, that is, only the fraction of the last relocation event (from the arrival point of the previous relocation to crossing the origin) is included in the computation. A first hitting event occurs when the end point of a jump arrives exactly at the origin. It is also called the first arrival event in physical literature. In the case of Lévy flight, the first hitting corresponds to an exact landing at the target coordinate. For Lévy walk, the first hitting event occurs when the last relocation ends at the target, that is, the searcher cannot identify the target while crossing it during the relocation event.

1.5.2.1 First passage time of Lévy flight and Lévy walk

The first passage property of Lévy flight, due to their Markovian character and the symmetric jump length distribution, is characterised by the Sparre Andersen-scaling [34, 155] in the long time limit. If the Lévy flight is described by the discrete Langevin equation

$$x_{n+1} - x_n = (K_\alpha \tau)^{1/\alpha} \xi_\alpha(n) \qquad (1.289)$$

with τ being the time step and $\xi_\alpha(n)$ the random variables sampled from a symmetric Lévy stable distribution with the characteristic function $\exp(-|k|^\alpha)$, the analytical expression of the limiting behavior of first passage time is [101]

$$p_{\mathrm{PF}}(t) \simeq \frac{x_0^{\alpha/2}}{\alpha\sqrt{\pi K_\alpha}\Gamma(\alpha/2)} t^{-3/2}, \qquad (1.290)$$

where x_0 is the initial position and α is the stable index of Lévy jumps. Lévy flight with smaller α has a higher propensity for long jumps, and the first passage behavior converges to the known Lévy-Smirnov law for Brownian motion when $\alpha \to 2$.

In contrast to Lévy flight, Lévy walk spatiotemporally couples each relocation distance x with a time cost $t = x/v_0$. Nevertheless, starting from a Lévy stable jump length distribution, the points of visitation of Lévy walk are identical with those of Lévy flight with the same entries for the sequence of jumps, with only the time counter for reaching the points of visitation differing for both processes. Hence, Lévy walk can be understood as Lévy flight with transformed durations of individual jumps. Therefore, for $\alpha > 1$, the average duration of a jump is finite, and thus in the limit of a large number of jumps, the time characteristics of Lévy walk and Lévy flight will only differ by a prefactor but should have the same scaling in time, i.e.,

$$p_{\mathrm{PW}}(t) \propto t^{-3/2}, \qquad \alpha > 1, \tag{1.291}$$

where $p_{\mathrm{PW}}(t)$ is the first passage time of Lévy walk.

For Lévy walk with $\alpha \leq 1$, the long time scaling of first passage time reads [15, 100, 145]

$$p_{\mathrm{PW}} \propto t^{-\alpha/2-1}, \qquad \alpha \leq 1. \tag{1.292}$$

This scaling form can be deduced from the argument of survival probability. In the case of a first-passage scenario, the survival probability of a Lévy flight is inversely proportional to the square root of the number n of flights, that is, $S_{\mathrm{LF}}(n) \propto n^{-1/2}$. For $\alpha < 1$, the number of jumps scales like $n \simeq t^{\alpha}$, such that $S_{\mathrm{LF}}(t) \propto t^{-\alpha/2}$. Since the first passage time PDF is the negative of the first derivative of the survival probability, we get the scaling form in Eq. (1.292).

1.5.2.2 First hitting time of Lévy flight and Lévy walk

In order to observe the first hitting events in one dimension, we need to require that the mean relocation time exists, corresponding to the requirement $\alpha > 1$ [146, 147]. For the opposite case of $0 < \alpha < 1$, the associated first hitting time PDF vanishes identically to zero. We do not consider the latter case here. For Lévy walk, the first hitting event occurs when the end point of a relocation with speed v_0 hits the target.

The probability of first hitting event clearly depends on the exact target size. Here we will concentrate on the case of point-like targets. The analytical derivations of the PDF $p_{\mathrm{HF}}(t)$ of the first hitting time of Lévy flight are based on the fractional Fokker-Planck equation with a sink term:

$$\frac{\partial p(x,t)}{\partial t} = K_\alpha \frac{\partial^\alpha p(x,t)}{\partial |x|^\alpha} - p_{\mathrm{HF}}(t)\delta(x). \qquad (1.293)$$

This equation can be solved in Laplace space $(t \to \lambda)$ for the initial condition $p(x,0) = \delta(x - x_0)$, i.e.,

$$\hat{p}_{\mathrm{HF}}(\lambda) = \int_{-\infty}^{\infty} \frac{e^{ikx_0}}{\lambda + K_\alpha |k|^\alpha} dk \left[\int_{-\infty}^{\infty} \frac{1}{\lambda + K_\alpha |k|^\alpha} dk \right]^{-1}. \qquad (1.294)$$

Its inverse is

$$p_{\mathrm{HF}}(t) = \frac{\alpha \sin(\pi/\alpha) K_\alpha^{1/\alpha} t^{1/\alpha-1}}{\pi} \int_0^\infty \cos(kx_0) E_{1,1/\alpha}(-K_\alpha k^\alpha t) dk, \qquad (1.295)$$

where $E_{\alpha,\beta}(z)$ is a two-parameter Mittag Leffler function defined by

$$E_{\alpha,\beta}(z) = \sum_{n=0}^{\infty} \frac{z^n}{\Gamma(\beta + n\alpha)}. \qquad (1.296)$$

By use of the series expansion of Mittag-Leffler function

$$E_{\alpha,\beta}(-z) = \sum_{n=1}^{\infty} \frac{(-1)^{n+1}}{z^n \Gamma(\beta - n\alpha)}, \qquad z > 0 \qquad (1.297)$$

around infinity, the $p_{\mathrm{HF}}(t)$ in Eq. (1.295) has the asymptotic form

$$p_{\mathrm{HF}} \propto t^{1/\alpha-2} \qquad (1.298)$$

for long time. This result recovers to $t^{-3/2}$ when $\alpha = 2$ for Brownian motion, which is consistent to the Lévy-Smirnov law of PDF of first passage time in Eq. (1.290).

The PDF p_{HW} of the first hitting time of Lévy walk shows the similar scaling exponents with the one of Lévy flight. The reason for this similarity is the same as for the case of first passage time: in the long time limit for $\alpha > 1$ both processes have the same scaling behavior due to the existence of a finite scale of the jumps in the processes. The only difference is that the prefactors in both cases may differ, corresponding to a renormalisation of the mean step time [145].

CHAPTER 2

Numerical Methods for the Governing Equations of PDF of Statistical Observables

From the previous chapters, it's easy to see that the statistical observables play an important role in describing the anomalous dynamics, say, position, first-passage time, mean exit time, and so on. Fractional Fokker-Planck equation is one of the most important equations of statistical physics, which describes the time evolution of the probability density function of positions of particles in anomalous diffusion. However, the solutions of the most of these equations can't be obtained explicitly and this fact motivates many authors to develop the effective numerical methods for these equations.

In the past few decades, various numerical methods [46, 47, 66, 79, 84–90, 108, 113, 116, 118, 133] have been proposed for the fractional Fokker-Planck equation, such as finite difference method, finite element method, and so on. As for finite difference method, [116] provides L_1 scheme to discretize time fractional derivatives and [87] gives optimal error estimates of L_1 scheme by comparing

DOI: 10.1201/9781003279099-2

the Laplace transforms of the exact solution and numerical solution; [88] uses the convolution quadrature method to propose the numerical scheme for fractional Fokker-Planck equations and [90] gives the k-th order scheme by modifying the first k-term of the numerical scheme. As for finite element method, [46] studies the finite element scheme for the fractional Fokker-Planck equation; [4] discusses the convergence of the finite element scheme for the Fokker-Planck equation involving fractional Laplacian operator and provides the optimal convergence rates for $s \in (\frac{1}{2}, 1)$.

With the deep insight on the mechanism of anomalous diffusion, in some cases, in order to model the real natural phenomena more accurately, the concept of internal states has to be introduced. Specifying each internal state with particular waiting time and jump length distributions and introducing a Markov chain with its transition matrix deciding the transition of the internal states, [189], recently, builds the multiple-internal-states fractional Fokker-Planck system (see [190] for the multiple-internal-states Lévy walk). In this chapter, we study the numerical method for the fractional Fokker-Planck system involving time fractional and time-space fractional systems.

The organization in this chapter is as follows. In Section 2.1, we introduce convolution quadrature method briefly and propose the temporal semi-discrete scheme for time fractional Fokker-Planck system. In Section 2.2, we first provide the regularity of the space-time fractional Fokker-Planck system, and then use the finite element method and L_1 method to discretize the spatial and temporal operator respectively and provide the optimal error estimates for the numerical scheme.

2.1 NUMERICAL METHODS FOR THE TIME FRACTIONAL FOKKER-PLANCK SYSTEM WITH TWO INTERNAL STATES

In this section, we, respectively, provide the time semi-discrete schemes based on the convolution quadrature generated by backward Euler and second-order backward difference for the following fractional Fokker-Planck system with two internal states [189, 190], i.e.,

$$\begin{cases} \mathbf{M}^T \dfrac{\partial}{\partial t}\mathbf{G} = (\mathbf{M}^T - \mathbf{I})\mathrm{diag}(\,_0D_t^{1-\alpha_1},\ _0D_t^{1-\alpha_2})\mathbf{G} \\[2mm] \qquad\quad + \mathbf{M}^T\mathrm{diag}(\,_0D_t^{1-\alpha_1},\ _0D_t^{1-\alpha_2})A\mathbf{G} \\[2mm] \qquad\quad + \mathbf{M}^T\mathbf{F} & \text{in } \Omega,\ t \in (0,T], \\[2mm] \mathbf{G}(\cdot,0) = \mathbf{G}_0 & \text{in } \Omega, \\[2mm] \mathbf{G} = 0 & \text{on } \partial\Omega,\ t \in (0,T], \end{cases}$$

$$(2.1)$$

where Ω denotes a bounded convex polygonal domain in $\mathbb{R}^d$ ($d = 1,2,3$); $A = -\Delta : H_0^1(\Omega) \bigcap H^2(\Omega) \to L^2(\Omega)$ is the negative Laplacian operator with a zero Dirichlet boundary condition; $\mathbf{M}$ is the transition matrix of a Markov chain and we assume it is a 2×2 invertible matrix, i.e.,

$$\mathbf{M} = \begin{bmatrix} m & 1-m \\ 1-m & m \end{bmatrix};$$

$\mathbf{M}^T$ denotes the transpose of $\mathbf{M}$; $\mathbf{G} = [G_1, G_2]^T$ denotes the solution of the system (2.1) and $\mathbf{F} = [f_1, f_2]^T$ is the source term; $\mathbf{G}_0 = [G_{1,0}, G_{2,0}]^T$ is the initial value; $\mathbf{I}$ is an identity matrix; T is a fixed final time; "diag" denotes a diagonal matrix formed from its vector argument, and $_0D_t^{1-\alpha_i}$, $\alpha_i \in (0,1)$, $i = 1,2$ are the Riemann-Liouville fractional derivatives defined by [150]

$$_0D_t^{1-\alpha_i}G = \frac{1}{\Gamma(\alpha_i)}\frac{\partial}{\partial t}\int_0^t (t-\xi)^{\alpha_i-1}G(\xi)d\xi, \quad \alpha_i \in (0,1). \quad (2.2)$$

2.1.1 Preliminaries

In the following, we abbreviate the functions $G_1(\cdot,t)$, $G_2(\cdot,t)$, $f_1(\cdot,t)$, and $f_2(\cdot,t)$ as $G_1(t)$, $G_2(t)$, $f_1(t)$, and $f_2(t)$, respectively. We denote $\|\cdot\|_{X\to Y}$ as the operator norm from X to Y with X and Y being Banach spaces. In particular, we denote $\|\cdot\|$ as the operator norm from $L^2(\Omega)$ to $L^2(\Omega)$. And the notation "~" means taking the Laplace transform. Define the following two sectors Σ_θ and $\Sigma_{\theta,\kappa}$ as, for $\kappa > 0$ and $\pi/2 < \theta < \pi$,

$$\Sigma_\theta = \{z \in \mathbb{C} : z \neq 0, |\arg z| \leq \theta\},$$
$$\Sigma_{\theta,\kappa} = \{z \in \mathbb{C} : |z| \geq \kappa, |\arg z| \leq \theta\},$$

and define the contour $\Gamma_{\theta,\kappa}$ by

$$\Gamma_{\theta,\kappa} = \{re^{-i\theta} : r \geq \kappa\} \cup \{\kappa e^{i\psi} : |\psi| \leq \theta\} \cup \{re^{i\theta} : r \geq \kappa\},$$

where the circular arc is oriented counterclockwise and the two rays are oriented with an increasing imaginary part and $\mathbf{i}^2 = -1$.

And then we recall the convolution quadrature introduced in [122, 123]. Here, we consider approximating the following convolution integral numerically

$$I(t) = \int_0^t f(t-s)g(s)ds.$$

By convolution properties, we can rewrite $I(t)$ as

$$I(t) = \int_0^t \frac{1}{2\pi i} \int_{\Gamma_{\theta,\kappa}} e^{zs}\tilde{f}(z)dzg(t-s)ds$$

$$= \frac{1}{2\pi i} \int_{\Gamma_{\theta,\kappa}} \tilde{f}(z) \int_0^t e^{zs}g(t-s)dsdz.$$

Using backward Euler method to approximate the following equation

$$\frac{dy}{dt} = zy + g \tag{2.3}$$

with $y(0) = 0$, we have

$$\frac{y_n - y_{n-1}}{\tau} = zy_n + g_n, \tag{2.4}$$

where τ is the step size, $t_n = n\tau$, $g_n = g(t_n)$ and y_n is the numerical approximation of $y(t_n)$. Multiplying ζ^n on both sides of (2.4) and summing n from 1 to ∞, one can get

$$(\delta_\tau(\zeta) - z)\sum_{n=1}^{\infty} y_n\zeta^n = \sum_{n=1}^{\infty} g_n\zeta^n,$$

where

$$\delta_\tau(\zeta) = \frac{1-\zeta}{\tau}.$$

Thus one has

$$\sum_{n=1}^{\infty} y_n\zeta^n = (\delta_\tau(\zeta) - z)^{-1}\sum_{n=1}^{\infty} g_n\zeta^n,$$

which leads to

$$\sum_{n=1}^{\infty} I_h(t_n)\zeta^n = \frac{1}{2\pi i} \int_{\Gamma_{\theta,\kappa}} \tilde{f}(z)(\delta_\tau(\zeta) - z)^{-1} \sum_{n=1}^{\infty} g_n \zeta^n dz$$

$$= \tilde{f}(\delta_\tau(\zeta)) \sum_{n=1}^{\infty} g_n \zeta^n.$$

Thus, we can get the integral formula generated by backward Euler formula

$$I(t_n) \approx I_h(t_n) = \sum_{i=0}^{n-1} w_i g_{n-i},$$

where

$$\tilde{f}(\delta_\tau(\zeta)) = \sum_{n=0}^{\infty} w_n \zeta^n.$$

Remark 2.1 *Here, if we use high-order backward difference formula or other formula to discretize (2.3), the corresponding numerical integral formula can be obtained similarly. For the details, see [122, 123].*

2.1.2 Equivalent Form of (2.1) and Some Useful Lemmas

Here we first provide equivalent form of Eq. (2.1), and then some useful lemmas are offered to help us to obtain the error estimates.

Multiplying $(\mathbf{M}^T)^{-1}$ on both sides of (2.1), we can get the following equivalent form

$$\begin{cases} \dfrac{\partial G_1}{\partial t} + a \, _0D_t^{1-\alpha_1} G_1 - \, _0D_t^{1-\alpha_1} \Delta G_1 \\ \qquad = a \, _0D_t^{1-\alpha_2} G_2 + f_1 & \text{in } \Omega, \ t \in (0, T], \\ \dfrac{\partial G_2}{\partial t} + a \, _0D_t^{1-\alpha_2} G_2 - \, _0D_t^{1-\alpha_2} \Delta G_2 & (2.5) \\ \qquad = a \, _0D_t^{1-\alpha_1} G_1 + f_2 & \text{in } \Omega, \ t \in (0, T], \\ \mathbf{G}(\cdot, 0) = \mathbf{G}_0 & \text{in } \Omega, \\ \mathbf{G} = 0 & \text{on } \partial\Omega, \ t \in (0, T], \end{cases}$$

where $a = \frac{1-m}{2m-1}$ and $m \in [0, \frac{1}{2}) \cup (\frac{1}{2}, 1]$.

According to the fact $\widetilde{_0D_t^\alpha u}(z) = z^\alpha \tilde{u}(z)$, $\alpha \in (0, 1)$ [150], the Laplace transform of (2.5) can be written as

$$\begin{aligned} z\tilde{G}_1 + az^{1-\alpha_1}\tilde{G}_1 + z^{1-\alpha_1}A\tilde{G}_1 &= az^{1-\alpha_2}\tilde{G}_2 + \tilde{f}_1 + G_{1,0}, \\ z\tilde{G}_2 + az^{1-\alpha_2}\tilde{G}_2 + z^{1-\alpha_2}A\tilde{G}_2 &= az^{1-\alpha_1}\tilde{G}_1 + \tilde{f}_2 + G_{2,0}, \end{aligned} \qquad (2.6)$$

which leads to

$$
\begin{aligned}
\tilde{G}_1 &= \left(H_{\alpha_1}(z) z^{\alpha_1 - 1} \tilde{f}_1 + a H(z) z^{\alpha_1 - 1} \tilde{f}_2 \right) \\
&\quad + \left(H_{\alpha_1}(z) z^{\alpha_1 - 1} G_{1,0} + a H(z) z^{\alpha_1 - 1} G_{2,0} \right), \\
\tilde{G}_2 &= \left(a H(z) z^{\alpha_2 - 1} \tilde{f}_1 + H_{\alpha_2}(z) z^{\alpha_2 - 1} \tilde{f}_2 \right) \\
&\quad + \left(a H(z) z^{\alpha_2 - 1} G_{1,0} + H_{\alpha_2}(z) z^{\alpha_2 - 1} G_{2,0} \right),
\end{aligned}
\tag{2.7}
$$

where

$$
H_{\alpha_1}(z) = H(z)(z^{\alpha_2} + a + A), \quad H_{\alpha_2}(z) = H(z)(z^{\alpha_1} + a + A),
\tag{2.8}
$$

and

$$
H(z) = \left((z^{\alpha_1} + a + A)(z^{\alpha_2} + a + A) - a^2 \right)^{-1}.
\tag{2.9}
$$

As for $H(z)$, $H_{\alpha_1}(z)$ and $H_{\alpha_2}(z)$, we have following properties, which will help us to obtain the optimal error estimates in time direction.

Lemma 2.1 *Assume $z \in \Sigma_{\theta,\kappa}$ and $\kappa > \max\{2|a|^{1/\alpha_1}, 2|a|^{1/\alpha_2}\}$. Then there hold*

$$
\left\| (z^{\alpha_1} + a + A)^{-1} \right\| \leq C |z|^{-\alpha_1}, \quad \left\| (z^{\alpha_2} + a + A)^{-1} \right\| \leq C |z|^{-\alpha_2},
$$

where C is a positive constant independent of κ.

Proof: The facts $\kappa > 2|a|^{1/\alpha_1}$ and $z \in \Sigma_{\theta,\kappa}$ show $z^{\alpha_1} + a \neq 0$. Let $z^{\alpha_1} = re^{i\phi}$, where $r \geq \kappa^{\alpha_1}$ and $|\phi| \leq \alpha_1 \theta$, $\theta \in (\pi/2, \pi)$. Then we get $z^{\alpha_1} + a = r\cos(\phi) + a + ir\sin(\phi)$.

(1). When $\phi > 0$ and $r\cos(\phi) + a < 0$, one has

$$
\frac{r\cos(\phi) + a}{r\sin(\phi)} = \frac{\cos(\phi) + a/r}{\sin(\phi)} \geq \frac{\cos(\phi) - 1}{\sin(\phi)} \geq C,
$$

which yields $\arg(z^{\alpha_1} + a) \leq \bar{\theta}$ for some $\bar{\theta} \in (\pi/2, \pi)$; as for $r\cos(\phi) + a > 0$, $\arg(z^{\alpha_1} + a) \leq \frac{\pi}{2}$ holds directly.

(2). When $\phi < 0$, we can get $\arg(z^{\alpha_1} + a) \leq \bar{\theta}$ for some $\bar{\theta} \in (\pi/2, \pi)$ similarly. Therefore, for $z \in \Sigma_{\theta,\kappa}$, we have $z^{\alpha_1} + a \in \Sigma_{\bar{\theta}}$ for some $\bar{\theta} \in (\pi/2, \pi)$.

Moreover, we obtain

$$
\left\| (z^{\alpha_1} + a + A)^{-1} \right\| \leq C |z^{\alpha_1} + a|^{-1} \quad \forall z \in \Sigma_{\theta,\kappa}
$$

by using the following resolvent estimate [124]

$$\left\|(z+A)^{-1}\right\| \le C|z|^{-1} \ \forall z \in \Sigma_\theta.$$

By the facts $|z| > 2|a|^{1/\alpha_1}$ and $\frac{|z^{\alpha_1}+a|^{-1}}{|z|^{-\alpha_1}} = \frac{|z|^{\alpha_1}}{|z^{\alpha_1}+a|} \le \frac{|z|^{\alpha_1}}{|z^{\alpha_1}-|a||} \le C$, the desired result can be got. Similarly, the second conclusion can be obtained. □

Lemma 2.2 *For $z \in \Sigma_{\theta,\kappa}$ and $\kappa > \max\{2|a|^{1/\alpha_1}, 2|a|^{1/\alpha_2}\}$, then*

$$\|H(z)\| \le C|z|^{-\alpha_1-\alpha_2}, \quad \|H_{\alpha_1}(z)\| \le C|z|^{-\alpha_1}, \quad \|H_{\alpha_2}(z)\| \le C|z|^{-\alpha_2}$$

hold, where $H(z)$, $H_{\alpha_1}(z)$, and $H_{\alpha_2}(z)$ are defined by (2.9) and (2.8); C is a positive constant independent of κ.

Proof: According to (2.9), we have

$$((z^{\alpha_1} + a + A)(z^{\alpha_2} + a + A) - a^2)u = v,$$

which leads to

$$u = ((z^{\alpha_1} + a + A)(z^{\alpha_2} + a + A))^{-1} v \\ + a^2 ((z^{\alpha_1} + a + A)(z^{\alpha_2} + a + A))^{-1} u. \tag{2.10}$$

Taking L_2 norm on both sides of (2.10) and using Lemma 2.1 imply

$$\|u\|_{L^2(\Omega)} \le C|z|^{-\alpha_1-\alpha_2}\|v\|_{L^2(\Omega)} + Ca^2|z|^{-\alpha_1-\alpha_2}\|u\|_{L^2(\Omega)}.$$

Choosing $\kappa > \max\{2|a|^{1/\alpha_1}, 2|a|^{1/\alpha_2}\}$ such that $Ca^2|z|^{-\alpha_1-\alpha_2} < 1/2$, we have

$$\|u\|_{L^2(\Omega)} \le C|z|^{-\alpha_1-\alpha_2}\|v\|_{L^2(\Omega)},$$

which leads to the desired estimate. Also, we can get the last two conclusions similarly. □

Lemma 2.3 *When $z \in \Sigma_{\theta,\kappa}$, where $\kappa > \max\{2|a|^{1/\alpha_1}, 2|a|^{1/\alpha_2}\}$, one has*

$$\|AH(z)\| \le C \min\{|z|^{-\alpha_1}, |z|^{-\alpha_2}\},$$
$$\|AH_{\alpha_1}(z)\| \le C, \quad \|AH_{\alpha_2}(z)\| \le C,$$

where C is a positive constant independent of κ.

Proof: According to the following equalities

$$AH(z) = (z^{\alpha_1} + a + A)H(z) - (z^{\alpha_1} + a)H(z)$$
$$= (z^{\alpha_2} + a + A)H(z) - (z^{\alpha_2} + a)H(z).$$

To estimate $AH(z)$, the terms $(z^{\alpha_1} + a + A)H(z)$, $(z^{\alpha_1} + a)H(z)$, $(z^{\alpha_2} + a + A)H(z)$ and $(z^{\alpha_2} + a)H(z)$ need to be estimated. As for $(z^{\alpha_1} + a + A)H(z)$, let

$$(z^{\alpha_2} + a + A)u - a^2(z^{\alpha_1} + a + A)^{-1}u = v,$$

which results in

$$u = (z^{\alpha_2} + a + A)^{-1}v + a^2((z^{\alpha_1} + a + A)(z^{\alpha_2} + a + A))^{-1}u.$$

Taking $\kappa > \max\{2|a|^{1/\alpha_1}, 2|a|^{1/\alpha_2}\}$ such that $Ca^2|z|^{-\alpha_1-\alpha_2} < 1/2$, one has

$$\|(z^{\alpha_1} + a + A)H(z)\| \leq C|z|^{-\alpha_2}. \tag{2.11}$$

Similarly, we have

$$\|(z^{\alpha_1} + a)H(z)\| \leq C|z|^{-\alpha_2},$$
$$\|(z^{\alpha_2} + a)H(z)\| \leq C|z|^{-\alpha_1},$$
$$\|(z^{\alpha_2} + a + A)H(z)\| \leq C|z|^{-\alpha_1}.$$

Analogously, we can get the last two conclusions. $\qquad\square$

According to (2.7) and convolution properties, we find G_1 and G_2 can be written as the following convolution form, i.e.,

$$G_1 = \int_0^t \mathcal{L}^{-1}(H_{\alpha_1}(z)z^{\alpha_1-1})(s)f_1(t-s)ds$$

$$+ a \int_0^t \mathcal{L}^{-1}(H(z)z^{\alpha_1-1})(s)f_2(t-s)ds$$

$$+ \mathcal{L}^{-1}(H_{\alpha_1}(z)z^{\alpha_1-1})(t)G_{1,0} + \mathcal{L}^{-1}(aH(z)z^{\alpha_1-1})(t)G_{2,0},$$

$$G_2 = a \int_0^t \mathcal{L}^{-1}(H(z)z^{\alpha_2-1})(s)f_1(t-s)ds$$

$$+ \int_0^t \mathcal{L}^{-1}(H_{\alpha_2}(z)z^{\alpha_2-1})(s)f_2(t-s)ds$$

$$+ \mathcal{L}^{-1}(aH(z)z^{\alpha_2-1})(t)G_{1,0} + \mathcal{L}^{-1}(H_{\alpha_2}(z)z^{\alpha_2-1})(t)G_{2,0},$$

where $\mathcal{L}^{-1}$ denotes inverse Laplace transform. So here, we consider using convolution quadrature to obtain the discrete scheme for (2.5).

2.1.3 First-Order Scheme and Error Analysis

Let the time step size $\tau = T/L$, $L \in \mathbb{N}$, $t_i = i\tau$, $i = 0, 1, \ldots, L$ and $0 = t_0 < t_1 < \cdots < t_L = T$. Using backward Euler convolution quadrature for the system (2.5), we obtain the temporal semi-discrete scheme

$$
\begin{cases}
\dfrac{G_1^n - G_1^{n-1}}{\tau} + a \displaystyle\sum_{i=0}^{n-1} d_i^{(1-\alpha_1)} G_1^{n-i} + \sum_{i=0}^{n-1} d_i^{(1-\alpha_1)} A G_1^{n-i} \\
\qquad\qquad\qquad = a \displaystyle\sum_{i=0}^{n-1} d_i^{(1-\alpha_2)} G_2^{n-i} + f_1^n, \\[2mm]
\dfrac{G_2^n - G_2^{n-1}}{\tau} + a \displaystyle\sum_{i=0}^{n-1} d_i^{(1-\alpha_2)} G_2^{n-i} + \sum_{i=0}^{n-1} d_i^{(1-\alpha_2)} A G_2^{n-i} \\
\qquad\qquad\qquad = a \displaystyle\sum_{i=0}^{n-1} d_i^{(1-\alpha_1)} G_1^{n-i} + f_2^n, \\[2mm]
G_1^0 = G_1(0), \\
G_2^0 = G_2(0),
\end{cases}
$$

$$\tag{2.12}$$

where

$$
\sum_{i=0}^{\infty} d_i^{(\alpha)} \zeta^i = (\delta_\tau(\zeta))^\alpha, \quad 0 < \alpha < 1, \tag{2.13}
$$

G_1^n, G_2^n are the numerical solutions of G_1, G_2 at time t_n and $f_1^n = f_1(t_n)$, $f_2^n = f_2(t_n)$.

Then we provide the error analysis for the homogeneous and inhomogeneous problems seperately.

2.1.3.1 Error estimates for the homogeneous problem

We first consider the error estimates for system (2.12) when $f_1(t) = f_2(t) = 0$. To get the solutions of the system (2.12), multiplying ζ^n and summing from 1 to ∞ for the both sides of the first two equations in (2.12) show

$$
\sum_{n=1}^{\infty} \frac{\zeta^n G_1^n - \zeta^n G_1^{n-1}}{\tau} + a \sum_{n=1}^{\infty} \sum_{i=0}^{n-1} d_i^{(1-\alpha_1)} \zeta^n G_1^{n-i}
$$

$$
+ \sum_{n=1}^{\infty} \sum_{i=0}^{n-1} d_i^{(1-\alpha_1)} \zeta^n A G_1^{n-i}
$$

$$
= a \sum_{n=1}^{\infty} \sum_{i=0}^{n-1} d_i^{(1-\alpha_2)} \zeta^n G_2^{n-i},
$$

$$
\sum_{n=1}^{\infty} \frac{\zeta^n G_2^n - \zeta^n G_2^{n-1}}{\tau} + a \sum_{n=1}^{\infty} \sum_{i=0}^{n-1} d_i^{(1-\alpha_2)} \zeta^n G_2^{n-i}
$$

$$
+ \sum_{n=1}^{\infty} \sum_{i=0}^{n-1} d_i^{(1-\alpha_2)} \zeta^n A G_2^{n-i}
$$

$$
= a \sum_{n=1}^{\infty} \sum_{i=0}^{n-1} d_i^{(1-\alpha_1)} \zeta^n G_1^{n-i},
$$

which results in, after simple calculations and using (2.13)

$$
\sum_{i=1}^{\infty} G_1^i \zeta^i = \frac{\zeta}{\tau} \left(H_{\alpha_1} \left(\frac{1-\zeta}{\tau} \right) \left(\frac{1-\zeta}{\tau} \right)^{\alpha_1 - 1} G_1(0) \right.
$$

$$
\left. + a H \left(\frac{1-\zeta}{\tau} \right) \left(\frac{1-\zeta}{\tau} \right)^{\alpha_1 - 1} G_2(0) \right),
$$

$$
\sum_{i=1}^{\infty} G_2^i \zeta^i = \frac{\zeta}{\tau} \left(a H \left(\frac{1-\zeta}{\tau} \right) \left(\frac{1-\zeta}{\tau} \right)^{\alpha_2 - 1} G_1(0) \right.
$$

$$
\left. + H_{\alpha_2} \left(\frac{1-\zeta}{\tau} \right) \left(\frac{1-\zeta}{\tau} \right)^{\alpha_2 - 1} G_2(0) \right),
$$

(2.14)

where H, H_{α_1}, and H_{α_2} are defined by (2.8) and (2.9).

Now we give the error estimates of the solutions of the systems (2.5) and (2.12) when $f_1 = 0$ and $f_2 = 0$.

Theorem 2.1 *Let G_1, G_2 and G_1^n, G_2^n be the solutions of the systems (2.5) and (2.12) with $G_{1,0}, G_{2,0} \in L^2(\Omega)$ and $f_1 = 0$, $f_2 = 0$.*

Then there hold

$$\|G_1(t_n) - G_1^n\|_{L^2(\Omega)} \le C\tau(t_n^{-1}\|G_{1,0}\|_{L^2(\Omega)} + t_n^{\alpha_2-1}\|G_{2,0}\|_{L^2(\Omega)}),$$
$$\|G_2(t_n) - G_2^n\|_{L^2(\Omega)} \le C\tau(t_n^{\alpha_1-1}\|G_{1,0}\|_{L^2(\Omega)} + t_n^{-1}\|G_{2,0}\|_{L^2(\Omega)}).$$

Proof: We first consider the error estimates between G_1^n and $G_1(t_n)$. By (2.14), for small $\xi_\tau = e^{-\tau(\kappa+1)}$, there is

$$G_1^n = \frac{1}{2\pi i \tau} \int_{|\zeta|=\xi_\tau} \zeta^{-n-1}\zeta \cdot \left(H_{\alpha_1}\left(\frac{1-\zeta}{\tau}\right)\left(\frac{1-\zeta}{\tau}\right)^{\alpha_1-1} G_1(0) \right.$$
$$\left. + aH\left(\frac{1-\zeta}{\tau}\right)\left(\frac{1-\zeta}{\tau}\right)^{\alpha_1-1} G_2(0) \right) d\zeta.$$

Taking $\zeta = e^{-z\tau}$, we obtain

$$G_1^n = \frac{1}{2\pi i} \int_{\Gamma^\tau} e^{zt_n} e^{-z\tau} \cdot \left(H_{\alpha_1}\left(\frac{1-e^{-z\tau}}{\tau}\right)\left(\frac{1-e^{-z\tau}}{\tau}\right)^{\alpha_1-1} G_1(0) \right.$$
$$\left. + aH\left(\frac{1-e^{-z\tau}}{\tau}\right)\left(\frac{1-e^{-z\tau}}{\tau}\right)^{\alpha_1-1} G_2(0) \right) dz,$$

where $\Gamma^\tau = \{z = \kappa + 1 + iy : y \in \mathbb{R} \text{ and } |y| \le \pi/\tau\}$. Deforming the contour Γ^τ to $\Gamma_{\theta,\kappa}^\tau = \{z \in \mathbb{C} : \kappa \le |z| \le \frac{\pi}{\tau\sin(\theta)}, |\arg z| = \theta\} \cup \{z \in \mathbb{C} : |z| = \kappa, |\arg z| \le \theta\}$, one has

$$G_1^n = \frac{1}{2\pi i} \int_{\Gamma_{\theta,\kappa}^\tau} e^{zt_n} e^{-z\tau} \cdot \left(H_{\alpha_1}\left(\frac{1-e^{-z\tau}}{\tau}\right)\left(\frac{1-e^{-z\tau}}{\tau}\right)^{\alpha_1-1} G_1(0) \right.$$
$$\left. + aH\left(\frac{1-e^{-z\tau}}{\tau}\right)\left(\frac{1-e^{-z\tau}}{\tau}\right)^{\alpha_1-1} G_2(0) \right) dz.$$

$$(2.15)$$

On the other hand, taking $f_1 = 0$, and $f_2 = 0$ and using (2.7), one has

$$\tilde{G}_1 = z^{\alpha_1-1}\left(H_{\alpha_1}(z)G_1(0) + aH(z)G_2(0)\right),$$

which gives

$$G_1(t) = \frac{1}{2\pi \mathbf{i}} \int_{\Gamma_{\theta,\kappa}} e^{zt} \left(H_{\alpha_1}(z) z^{\alpha_1-1} G_1(0) + aH(z) z^{\alpha_1-1} G_2(0) \right) dz.$$
(2.16)

Subtracting (2.15) from (2.16) leads to

$$G_1(t_n) - G_1^n$$
$$= \frac{1}{2\pi \mathbf{i}} \int_{\Gamma_{\theta,\kappa} \backslash \Gamma_{\theta,\kappa}^\tau} e^{zt_n} H_{\alpha_1}(z) z^{\alpha_1-1} G_1(0) dz$$
$$+ \frac{1}{2\pi \mathbf{i}} \int_{\Gamma_{\theta,\kappa} \backslash \Gamma_{\theta,\kappa}^\tau} e^{zt_n} aH(z) z^{\alpha_1-1} G_2(0) dz$$
$$+ \frac{1}{2\pi \mathbf{i}} \int_{\Gamma_{\theta,\kappa}^\tau} e^{zt_n} \left(H_{\alpha_1}(z) z^{\alpha_1-1} - e^{-z\tau} H_{\alpha_1} \left(\frac{1-e^{-z\tau}}{\tau} \right) \right.$$
$$\left. \cdot \left(\frac{1-e^{-z\tau}}{\tau} \right)^{\alpha_1-1} \right) G_1(0) dz$$
$$+ \frac{a}{2\pi \mathbf{i}} \int_{\Gamma_{\theta,\kappa}^\tau} e^{zt_n} \left(H(z) z^{\alpha_1-1} - e^{-z\tau} H \left(\frac{1-e^{-z\tau}}{\tau} \right) \right.$$
$$\left. \cdot \left(\frac{1-e^{-z\tau}}{\tau} \right)^{\alpha_1-1} \right) G_2(0) dz = I + II + III + IV.$$

Using Lemma 2.2, we have

$$\|I\|_{L^2(\Omega)} \le C \int_{\Gamma_{\theta,\kappa} \backslash \Gamma_{\theta,\kappa}^\tau} e^{-C|z|t_n} |z|^{\alpha_1-1} \|H_{\alpha_1}(z)\| |dz| \|G_1(0)\|_{L^2(\Omega)}$$
$$\le Ct_n^{-1} \tau \|G_1(0)\|_{L^2(\Omega)}.$$

Similarly, one has

$$\|II\|_{L^2(\Omega)} \le C \int_{\Gamma_{\theta,\kappa} \backslash \Gamma_{\theta,\kappa}^\tau} e^{-C|z|t_n} |z|^{\alpha_1-1} \|H(z)\| |dz| \|G_2(0)\|_{L^2(\Omega)}$$
$$\le Ct_n^{\alpha_2-1} \tau \|G_2(0)\|_{L^2(\Omega)}.$$

Next for III and IV, we obtain

$$III = \frac{1}{2\pi i} \int_{\Gamma_{\theta,\kappa}^{\tau}} e^{zt_n} e^{-z\tau} \left(e^{z\tau} H_{\alpha_1}(z) z^{\alpha_1 - 1} - H_{\alpha_1} \left(\frac{1 - e^{-z\tau}}{\tau} \right) \right.$$

$$\left. \cdot \left(\frac{1 - e^{-z\tau}}{\tau} \right)^{\alpha_1 - 1} \right) G_1(0) dz$$

$$= \frac{1}{2\pi i} \int_{\Gamma_{\theta,\kappa}^{\tau}} e^{zt_n} e^{-z\tau}$$

$$\cdot \left(H_{\alpha_1}(z) z^{\alpha_1 - 1} - H_{\alpha_1} \left(\frac{1 - e^{-z\tau}}{\tau} \right) \left(\frac{1 - e^{-z\tau}}{\tau} \right)^{\alpha_1 - 1} \right)$$

$$\cdot G_1(0) dz + \frac{1}{2\pi i} \int_{\Gamma_{\theta,\kappa}^{\tau}} e^{zt_n} e^{-z\tau} (e^{z\tau} - 1) H_{\alpha_1}(z) z^{\alpha_1 - 1} G_1(0) dz$$

$$= III_1 + III_2$$

and

$$IV = \frac{a}{2\pi i} \int_{\Gamma_{\theta,\kappa}^{\tau}} e^{zt_n} e^{-z\tau}$$

$$\cdot \left(e^{z\tau} H(z) z^{\alpha_1 - 1} - H \left(\frac{1 - e^{-z\tau}}{\tau} \right) \left(\frac{1 - e^{-z\tau}}{\tau} \right)^{\alpha_1 - 1} \right)$$

$$\cdot G_2(0) dz$$

$$= \frac{a}{2\pi i} \int_{\Gamma_{\theta,\kappa}^{\tau}} e^{zt_n} e^{-z\tau}$$

$$\cdot \left(H(z) z^{\alpha_1 - 1} - H \left(\frac{1 - e^{-z\tau}}{\tau} \right) \left(\frac{1 - e^{-z\tau}}{\tau} \right)^{\alpha_1 - 1} \right)$$

$$\cdot G_2(0) dz + \frac{a}{2\pi i} \int_{\Gamma_{\theta,\kappa}^{\tau}} e^{zt_n} e^{-z\tau} (e^{z\tau} - 1) H(z) z^{\alpha_1 - 1} G_2(0) dz$$

$$= IV_1 + IV_2.$$

Using the properties $\| \frac{d}{dz}(H_{\alpha_1}(z) z^{\alpha_1 - 1}) \| \leq C|z|^{-2}$, $\| \frac{d}{dz}(H(z) z^{\alpha_1 - 1}) \| \leq C|z|^{-\alpha_2 - 2}$, the mean value theorem, and the fact $\left(\frac{1 - e^{-z\tau}}{\tau} \right) = z + \mathcal{O}(\tau z^2)$, the following estimates

$$\left\| H_{\alpha_1}(z) z^{\alpha_1 - 1} - H_{\alpha_1} \left(\frac{1 - e^{-z\tau}}{\tau} \right) \left(\frac{1 - e^{-z\tau}}{\tau} \right)^{\alpha_1 - 1} \right\| \leq C\tau$$

and

$$\left\| H(z)z^{\alpha_1-1} - H\left(\frac{1-e^{-z\tau}}{\tau}\right)\left(\frac{1-e^{-z\tau}}{\tau}\right)^{\alpha_1-1} \right\| \leq C\tau|z|^{-\alpha_2}$$

can be obtained. Thus one can get

$$\|III_1\|_{L^2(\Omega)} \leq C\tau \int_{\Gamma_{\theta,\kappa}^\tau} e^{Re(z)t_{n-1}}|dz| \|G_1(0)\|_{L^2(\Omega)}$$

$$\leq Ct_n^{-1}\tau\|G_1(0)\|_{L^2(\Omega)}$$

and

$$\|IV_1\|_{L^2(\Omega)} \leq C\tau \int_{\Gamma_{\theta,\kappa}^\tau} e^{Re(z)t_{n-1}}|z|^{-\alpha_2}|dz| \|G_2(0)\|_{L^2(\Omega)}$$

$$\leq Ct_n^{\alpha_2-1}\tau\|G_2(0)\|_{L^2(\Omega)},$$

where $Re(z)$ means the real part of z. The fact $|e^{z\tau} - 1| \leq C\tau|z|$ and Lemma 2.2 imply

$$\|III_2\|_{L^2(\Omega)} \leq C\tau \int_{\Gamma_{\theta,\kappa}^\tau} e^{Re(z)t_{n-1}}|dz| \|G_1(0)\|_{L^2(\Omega)}$$

$$\leq Ct_n^{-1}\tau\|G_1(0)\|_{L^2(\Omega)},$$

$$\|IV_2\|_{L^2(\Omega)} \leq C\tau \int_{\Gamma_{\theta,\kappa}^\tau} e^{Re(z)t_{n-1}}|z|^{-\alpha_2}|dz| \|G_2(0)\|_{L^2(\Omega)}$$

$$\leq Ct_n^{\alpha_2-1}\tau\|G_2(0)\|_{L^2(\Omega)}.$$

In summary,

$$\|G_1(t_n) - G_1^n\|_{L^2(\Omega)} \leq C\tau(t_n^{-1}\|G_1(0)\|_{L^2(\Omega)} + t_n^{\alpha_2-1}\|G_2(0)\|_{L^2(\Omega)}).$$

Analogously, we have

$$\|G_2(t_n) - G_2^n\|_{L^2(\Omega)} \leq C\tau(t_n^{\alpha_1-1}\|G_1(0)\|_{L^2(\Omega)} + t_n^{-1}\|G_2(0)\|_{L^2(\Omega)}).$$

So we complete the proof. □

2.1.3.2 *Error estimates for the inhomogeneous problem*

Now we consider the error estimates for the inhomogeneous problem with vanishing initial value. Similar to the derivation of (2.14), one has

$$\sum_{i=0}^{\infty} G_1^i \zeta^i = H_{\alpha_1} \left(\frac{1-\zeta}{\tau} \right) \left(\frac{1-\zeta}{\tau} \right)^{\alpha_1-1} \sum_{i=1}^{\infty} \zeta^i f_1^i$$

$$+ aH \left(\frac{1-\zeta}{\tau} \right) \left(\frac{1-\zeta}{\tau} \right)^{\alpha_1-1} \sum_{i=1}^{\infty} \zeta^i f_2^i,$$

$$\sum_{i=0}^{\infty} G_2^i \zeta^i = aH \left(\frac{1-\zeta}{\tau} \right) \left(\frac{1-\zeta}{\tau} \right)^{\alpha_2-1} \sum_{i=1}^{\infty} \zeta^i f_1^i$$

$$+ H_{\alpha_2} \left(\frac{1-\zeta}{\tau} \right) \left(\frac{1-\zeta}{\tau} \right)^{\alpha_2-1} \sum_{i=1}^{\infty} \zeta^i f_2^i.$$

(2.17)

Then we have the error estimates of the solutions of the systems (2.5) and (2.12) when $G_{1,0} = 0$, $G_{2,0} = 0$.

Theorem 2.2 *Let G_1, G_2 and G_1^n, G_2^n be the solutions of the systems (2.2) and (2.12) with $G_{1,0} = 0$, $G_{2,0} = 0$, $f_1(0), f_2(0) \in L^2(\Omega)$ and $\int_0^{t_n} \|f_1'(s)\|_{L^2(\Omega)} ds < \infty$, $\int_0^{t_n} \|f_2'(s)\|_{L^2(\Omega)} ds < \infty$, where $f_i'(t)$ $(i = 1, 2)$ denote the first derivative about t. Then*

$$\|G_1(t_n) - G_1^n\|_{L^2(\Omega)} \leq C\tau \left(\int_0^{t_n} \|f_1'(s)\|_{L^2(\Omega)} ds + \|f_1(0)\|_{L^2(\Omega)} \right.$$

$$+ \int_0^{t_n} \|f_2'(s)\|_{L^2(\Omega)} ds + \|f_2(0)\|_{L^2(\Omega)} \right),$$

$$\|G_2(t_n) - G_2^n\|_{L^2(\Omega)} \leq C\tau \left(\int_0^{t_n} \|f_1'(s)\|_{L^2(\Omega)} ds + \|f_1(0)\|_{L^2(\Omega)} \right.$$

$$+ \int_0^{t_n} \|f_2'(s)\|_{L^2(\Omega)} ds + \|f_2(0)\|_{L^2(\Omega)} \right).$$

Proof: We mainly provide the error estimate between G_1^n and $G_1(t_n)$ in detail. Introduce $E_{1,j}$ and $E_{\alpha_1,j}$ satisfying

$$aH \left(\frac{1-\zeta}{\tau} \right) \left(\frac{1-\zeta}{\tau} \right)^{\alpha_1-1} = \sum_{j=0}^{\infty} E_{1,j} \zeta^j$$

and

$$H_{\alpha_1} \left(\frac{1-\zeta}{\tau} \right) \left(\frac{1-\zeta}{\tau} \right)^{\alpha_1-1} = \sum_{j=0}^{\infty} E_{\alpha_1,j} \zeta^j.$$

Then we can rewrite (2.17) as

$$\sum_{i=0}^{\infty} G_1^i \zeta^i = \left(\sum_{j=0}^{\infty} E_{\alpha_1,j} \zeta^j \right) \left(\sum_{i=1}^{\infty} \zeta^i f_1^i \right) + \left(\sum_{j=0}^{\infty} E_{1,j} \zeta^j \right) \left(\sum_{i=1}^{\infty} \zeta^i f_2^i \right).$$

Thus

$$G_1^n = \lim_{t \to t_n^-} \left(\left(\hat{E}_{\alpha_1} * f_1 \right)(t) + \left(\hat{E}_1 * f_2 \right)(t) \right), \tag{2.18}$$

where $\hat{E}_{\alpha_1} = \sum_{j=0}^{\infty} E_{\alpha_1,j}\delta_{t_j}$, $\hat{E}_1 = \sum_{j=0}^{\infty} E_{1,j}\delta_{t_j}$ with δ_t the delta function concentrated at t and '$*$' denotes the convolution. Introducing

$$Q_{\alpha_1}(t',v) = \hat{E}_{\alpha_1} * v, \quad Q_1(t',v) = \hat{E}_1 * v,$$

we have

$$\begin{aligned}
G_1^n = &\lim_{t \to t_n^-} \left((Q_{\alpha_1}(t',1) * f_1')(t) + (Q_1(t',1) * f_2')(t) \right) \\
&+ \lim_{t \to t_n^-} \left(Q_{\alpha_1}(t,1)f_1(0) + Q_1(t,1)f_2(0) \right),
\end{aligned} \tag{2.19}$$

where we use the fact $f_1(t) = f_1(0)+1*f_1'(t)$, $f_2(t) = f_2(0)+1*f_2'(t)$ and (2.18). Similarly,

$$\begin{aligned}
G_1(t) = &\frac{1}{2\pi\mathbf{i}} \left(\int_{\Gamma_{\theta,\kappa}} e^{zt'} H_{\alpha_1}(z)z^{\alpha_1-1}z^{-1}dz * f_1'(t') \right)(t) \\
&+ \frac{a}{2\pi\mathbf{i}} \left(\int_{\Gamma_{\theta,\kappa}} e^{zt'} H(z)z^{\alpha_1-1}z^{-1}dz * f_2'(t') \right)(t) \\
&+ \frac{1}{2\pi\mathbf{i}} \int_{\Gamma_{\theta,\kappa}} e^{zt} H_{\alpha_1}(z)z^{\alpha_1-1}z^{-1}dz f_1(0) \\
&+ \frac{a}{2\pi\mathbf{i}} \int_{\Gamma_{\theta,\kappa}} e^{zt} H(z)z^{\alpha_1-1}z^{-1}dz f_2(0)
\end{aligned} \tag{2.20}$$

can be obtained from (2.7). Subtracting (2.19) from (2.20) results in

$$\begin{aligned}
G_1(t_n) - G_1^n = &\lim_{t \to t_n^-} \left((I(t') * f_1'(t'))(t) + (II(t') * f_2'(t'))(t) \right) \\
&+ \lim_{t \to t_n^-} \left(I(t)f_1(0) + II(t)f_2(0) \right),
\end{aligned}$$

where

$$I(t) = \frac{1}{2\pi\mathbf{i}} \int_{\Gamma_{\theta,\kappa}} e^{zt} H_{\alpha_1}(z)z^{\alpha_1-1}z^{-1}dz - Q_{\alpha_1}(t,1),$$

$$II(t) = \frac{a}{2\pi\mathbf{i}} \int_{\Gamma_{\theta,\kappa}} e^{zt} H(z)z^{\alpha_1-1}z^{-1}dz - Q_1(t,1).$$

For $t \in [t_0, t_1)$, the following estimation holds directly

$$\|I(t)\| \leq C\tau.$$

As for $t \in [t_{n-1}, t_n)$ with $n > 1$, taking κ large enough to satisfy the conditions in Lemma 2.2 (κ depends on $|a|$ and is independent of t), we have

$$\|I(t)\| = \left\| \frac{1}{2\pi i} \int_{\Gamma_{\theta,\kappa}} e^{zt} H_{\alpha_1}(z) z^{\alpha_1-1} z^{-1} dz \right.$$
$$- \frac{1}{2\pi i} \int_{\Gamma_{\theta,\kappa}} e^{zt_{n-1}} H_{\alpha_1}(z) z^{\alpha_1-1} z^{-1} dz \Big\|$$
$$+ \left\| \frac{1}{2\pi i} \int_{\Gamma_{\theta,\kappa}} e^{zt_{n-1}} H_{\alpha_1}(z) z^{\alpha_1-1} z^{-1} dz - Q_{\alpha_1}(t,1) \right\|$$
$$\leq C\tau + \left\| \frac{1}{2\pi i} \int_{\Gamma_{\theta,\kappa}} e^{zt_{n-1}} H_{\alpha_1}(z) z^{\alpha_1-1} z^{-1} dz - Q_{\alpha_1}(t,1) \right\|.$$

Choosing $\xi_\tau = e^{-\tau(\kappa+1)}$ gives

$$E_{\alpha_1,n} = \frac{1}{2\pi i} \int_{|\zeta|=\xi_\tau} \zeta^{-n-1} H_{\alpha_1} \left(\frac{1-\zeta}{\tau} \right) \left(\frac{1-\zeta}{\tau} \right)^{\alpha_1-1} d\zeta.$$

Then

$$Q_{\alpha_1}(t,1) = \sum_{j=0}^{n-1} E_{\alpha_1,j}$$
$$= \frac{1}{2\pi\tau i} \int_{|\zeta|=\xi_\tau} \zeta^{-n} H_{\alpha_1} \left(\frac{1-\zeta}{\tau} \right) \left(\frac{1-\zeta}{\tau} \right)^{\alpha_1-2} d\zeta,$$

where we use the fact $\sum_{j=0}^{n-1} \zeta^{-j-1} = (\zeta^{-n} - 1)/(1 - \zeta)$ and for small ζ, the term

$$\left(\frac{1-\zeta}{\tau} \right)^{\alpha_1-1} H_{\alpha_1} \left(\frac{1-\zeta}{\tau} \right) \frac{1}{1-\zeta}$$

is analytic.

Taking $\zeta = e^{-z\tau}$, one has

$$Q_{\alpha_1}(t,1) = \frac{1}{2\pi i} \int_{\Gamma^\tau} e^{zt_{n-1}} H_{\alpha_1} \left(\frac{1-e^{-z\tau}}{\tau} \right) \left(\frac{1-e^{-z\tau}}{\tau} \right)^{\alpha_1-2} dz,$$

where $\Gamma^\tau = \{z = \kappa + 1 + iy : y \in \mathbb{R} \text{ and } |y| \leq \pi/\tau\}$. Next deforming the contour Γ^τ to $\Gamma_{\theta,\kappa}^\tau = \{z \in \mathbb{C} : \kappa \leq |z| \leq \frac{\pi}{\tau \sin(\theta)}, |\arg z| = \theta\} \cup \{z \in \mathbb{C} : |z| = \kappa, |\arg z| \leq \theta\}$ leads to

$$Q_{\alpha_1}(t,1) = \frac{1}{2\pi i} \int_{\Gamma_{\theta,\kappa}^\tau} e^{zt_{n-1}} H_{\alpha_1} \left(\frac{1 - e^{-z\tau}}{\tau} \right) \left(\frac{1 - e^{-z\tau}}{\tau} \right)^{\alpha_1 - 2} dz,$$

which results in

$$\frac{1}{2\pi i} \int_{\Gamma_{\theta,\kappa}} e^{zt_{n-1}} H_{\alpha_1}(z) z^{\alpha_1 - 1} z^{-1} dz - Q_{\alpha_1}(t,1)$$

$$= \frac{1}{2\pi i} \int_{\Gamma_{\theta,\kappa} \backslash \Gamma_{\theta,\kappa}^\tau} e^{zt_{n-1}} H_{\alpha_1}(z) z^{\alpha_1 - 2} dz$$

$$+ \frac{1}{2\pi i} \int_{\Gamma_{\theta,\kappa}^\tau} e^{zt_{n-1}}$$

$$\cdot \left(H_{\alpha_1}(z) z^{\alpha_1 - 2} - H_{\alpha_1} \left(\frac{1 - e^{-z\tau}}{\tau} \right) \left(\frac{1 - e^{-z\tau}}{\tau} \right)^{\alpha_1 - 2} \right) dz$$

$$= I_1 + I_2.$$

For I_1, Lemma 2.2 yields

$$\|I_1\| \leq C \int_{\Gamma_{\theta,\kappa} \backslash \Gamma_{\theta,\kappa}^\tau} e^{-C|z|t_{n-1}} |z|^{-2} |dz| \leq C\tau.$$

For I_2, by $\left\| \frac{d}{dz} \left(H_{\alpha_1}(z) z^{\alpha_1 - 2} \right) \right\| \leq C|z|^{-3}$, the mean value theorem, and the fact $\left(\frac{1 - e^{-z\tau}}{\tau} \right) = z + \mathcal{O}(\tau z^2)$, we get

$$\left\| H_{\alpha_1}(z) z^{\alpha_1 - 2} - H_{\alpha_1} \left(\frac{1 - e^{-z\tau}}{\tau} \right) \left(\frac{1 - e^{-z\tau}}{\tau} \right)^{\alpha_1 - 2} \right\|$$

$$\leq C|z|^{-3} |\tau z^2| \leq C\tau |z|^{-1}.$$

Thus

$$\|I_2\| \leq C\tau \int_{\Gamma_{\theta,\kappa}^\tau} e^{Re(z)t_{n-1}} |z|^{-1} |dz| \leq C\tau,$$

and
$$\|I(t)\| \leq C\tau.$$

Similar to the derivation of $\|I(t)\|$, we can get $\|II(t)\| \leq C\tau$.

In summary, one has

$$
\|G_1(t_n) - G_1^n\|_{L^2(\Omega)} \leq C\tau \left((1 * \|f_1'\|_{L^2(\Omega)})(t_n) + (1 * \|f_2'\|_{L^2(\Omega)})(t_n) \right.
$$
$$
\left. + \|f_1(0)\|_{L^2(\Omega)} + \|f_2(0)\|_{L^2(\Omega)} \right).
$$

Also, there holds

$$
\|G_2(t_n) - G_2^n\|_{L^2(\Omega)} \leq C\tau \left((1 * \|f_1'\|_{L^2(\Omega)})(t_n) + (1 * \|f_2'\|_{L^2(\Omega)})(t_n) \right.
$$
$$
\left. + \|f_1(0)\|_{L^2(\Omega)} + \|f_2(0)\|_{L^2(\Omega)} \right).
$$

$$\square$$

2.1.4 Second-Order Scheme and Error Analysis

In this section, we provide the second-order backward difference (SBD) scheme for the homogeneous problem (2.5), i.e., $f_1 = f_2 = 0$, and give the error estimate of the SBD scheme. From [88, 122], we can only get first-order accuracy for $G_1(0) \neq 0$ and $G_2(0) \neq 0$ if we use the SBD convolution quadrature directly. According to [88,124], to keep the accuracy of the scheme, the discretization need to be modified. Namely, introducing $\bar{\partial}_\tau^\alpha$ and ∂_t^{-1} as the discretization of $_0D_t^\alpha$ and the integration on time respectively, we can get

$$
\bar{\partial}_\tau G_1^n + a\bar{\partial}_\tau^{1-\alpha_1}(G_1^n - G_1(0)) + \bar{\partial}_\tau^{1-\alpha_1}A(G_1^n - G_1(0))
$$
$$
- a\bar{\partial}_\tau^{1-\alpha_2}(G_2^n - G_2(0))
$$
$$
= -a\bar{\partial}_\tau^{1-\alpha_1}\bar{\partial}_\tau\partial_t^{-1}G_1(0) - \bar{\partial}_\tau^{1-\alpha_1}\bar{\partial}_\tau\partial_t^{-1}AG_1(0)
$$
$$
+ a\bar{\partial}_\tau^{1-\alpha_2}\bar{\partial}_\tau\partial_t^{-1}G_2(0),
$$
$$
\bar{\partial}_\tau G_2^n + a\bar{\partial}_\tau^{1-\alpha_2}(G_2^n - G_2(0)) + \bar{\partial}_\tau^{1-\alpha_2}A(G_2^n - G_2(0))
$$
$$
- a\bar{\partial}_\tau^{1-\alpha_1}(G_1^n - G_1(0))
$$
$$
= -a\bar{\partial}_\tau^{1-\alpha_2}\bar{\partial}_\tau\partial_t^{-1}G_2(0) - \bar{\partial}_\tau^{1-\alpha_2}\bar{\partial}_\tau\partial_t^{-1}AG_2(0)
$$
$$
+ a\bar{\partial}_\tau^{1-\alpha_1}\bar{\partial}_\tau\partial_t^{-1}G_1(0).
$$

According to the fact $(0, 3/2, 1, 1, \dots) = \bar{\partial}_\tau \partial_t^{-1} 1$ [88], the SBD scheme of the homogeneous problem (2.5) can be written as

$$
\begin{cases}
\dfrac{G_1^1 - G_1^0}{\tau} + a d_{2,0}^{1-\alpha_1} \left(\dfrac{2}{3} G_1^1 + \dfrac{1}{3} G_1^0 \right) + d_{2,0}^{1-\alpha_1} A \left(\dfrac{2}{3} G_1^1 + \dfrac{1}{3} G_1^0 \right) \\
\qquad\qquad\qquad = a d_{2,0}^{1-\alpha_2} \left(\dfrac{2}{3} G_2^1 + \dfrac{1}{3} G_2^0 \right) \quad \text{in } \Omega, \\[2ex]
\dfrac{G_2^1 - G_2^0}{\tau} + a d_{2,0}^{1-\alpha_2} \left(\dfrac{2}{3} G_2^1 + \dfrac{1}{3} G_2^0 \right) + d_{2,0}^{1-\alpha_2} A \left(\dfrac{2}{3} G_2^1 + \dfrac{1}{3} G_2^0 \right) \\
\qquad\qquad\qquad = a d_{2,0}^{1-\alpha_1} \left(\dfrac{2}{3} G_1^1 + \dfrac{1}{3} G_1^0 \right) \quad \text{in } \Omega, \\[2ex]
\dfrac{1}{\tau} \left(\dfrac{3}{2} G_1^n - 2 G_1^{n-1} + \dfrac{1}{2} G_1^{n-2} \right) + a \displaystyle\sum_{i=0}^{n-1} d_{2,i}^{1-\alpha_1} G_1^{n-i} + a \dfrac{1}{2} d_{2,n-1}^{1-\alpha_1} G_1^0 \\[1ex]
\qquad + \displaystyle\sum_{i=0}^{n-1} d_{2,i}^{1-\alpha_1} A G_1^{n-i} + \dfrac{1}{2} d_{2,n-1}^{1-\alpha_1} A G_1^0 \\[1ex]
\qquad = a \displaystyle\sum_{i=0}^{n-1} d_{2,i}^{1-\alpha_2} G_2^{n-i} + \dfrac{1}{2} a d_{2,n-1}^{1-\alpha_2} G_2^0 \qquad \text{in } \Omega, \ n \geq 2, \\[2ex]
\dfrac{1}{\tau} \left(\dfrac{3}{2} G_2^n - 2 G_2^{n-1} + \dfrac{1}{2} G_2^{n-2} \right) + a \displaystyle\sum_{i=0}^{n-1} d_{2,i}^{1-\alpha_2} G_2^{n-i} + a \dfrac{1}{2} d_{2,n-1}^{1-\alpha_2} G_2^0 \\[1ex]
\qquad + \displaystyle\sum_{i=0}^{n-1} d_{2,i}^{1-\alpha_2} A G_2^{n-i} + \dfrac{1}{2} d_{2,n-1}^{1-\alpha_2} A G_2^0 \\[1ex]
\qquad = a \displaystyle\sum_{i=0}^{n-1} d_{2,i}^{1-\alpha_1} G_1^{n-i} + \dfrac{1}{2} a d_{2,n-1}^{1-\alpha_1} G_1^0 \qquad \text{in } \Omega, \ n \geq 2, \\[2ex]
G_1^0 = G_1(0), \quad G_2^0 = G_2(0) \qquad\qquad \text{in } \Omega, \\[1ex]
G_1^n = G_2^n = 0 \qquad\qquad\qquad\qquad\quad \text{on } \partial\Omega, \ n \geq 0,
\end{cases}
\tag{2.21}
$$

where G_1^n, G_2^n are the numerical solutions of G_1, G_2 at t_n and $G_1^{-1} = G_1(0)$, $G_2^{-1} = G_2(0)$.

Next, we provide the error estimates between the homogeneous problem (2.5) and (2.21).

2.1.4.1 Error analysis

Theorem 2.3 *Let G_1, G_2 and G_1^n, G_2^n be, respectively, the solutions of the homogeneous problem* (2.5) *and* (2.21). *Then*

$$\|G_1(t_n) - G_1^n\|_{L^2(\Omega)}$$
$$\leq C\tau^2 \left(t_n^{-2}\|G_1(0)\|_{L^2(\Omega)} + t_n^{\alpha_2-2}\|G_2(0)\|_{L^2(\Omega)} \right),$$
$$\|G_2(t_n) - G_2^n\|_{L^2(\Omega)}$$
$$\leq C\tau^2 \left(t_n^{\alpha_1-2}\|G_1(0)\|_{L^2(\Omega)} + t_n^{-2}\|G_2(0)\|_{L^2(\Omega)} \right).$$

Proof: Introduce $\mathbf{V} = [V_1, V_2]^T$ with $V_1(t) = G_1(t) - G_1(0)$, $V_2(t) = G_2(t) - G_2(0)$ and denote $\bar{V}_1^n = G_1^n - G_1(0)$, $\bar{V}_2^n = G_2^n - G_2(0)$. We rewrite (2.5) as

$$
\begin{cases}
\dfrac{\partial V_1}{\partial t} + a\, {_0D_t^{1-\alpha_1}}V_1 + {_0D_t^{1-\alpha_1}}AV_1 - a\, {_0D_t^{1-\alpha_2}}V_2 = \\
\qquad - a\, {_0D_t^{1-\alpha_1}}G_1(0) - {_0D_t^{1-\alpha_1}}AG_1(0) \\
\qquad + a\, {_0D_t^{1-\alpha_2}}G_2(0) \qquad\qquad \text{in } \Omega,\ t \in [0,T], \\[2ex]
\dfrac{\partial V_2}{\partial t} + a\, {_0D_t^{1-\alpha_2}}V_2 + {_0D_t^{1-\alpha_2}}AV_2 - a\, {_0D_t^{1-\alpha_1}}V_1 = \\
\qquad - a\, {_0D_t^{1-\alpha_2}}G_2(0) - {_0D_t^{1-\alpha_2}}AG_2(0) \\
\qquad + a\, {_0D_t^{1-\alpha_1}}G_1(0) \qquad\qquad \text{in } \Omega,\ t \in [0,T], \\[2ex]
\mathbf{V}(\cdot,0) = 0 \qquad\qquad\qquad\qquad \text{in } \Omega, \\[1ex]
\mathbf{V} = 0 \qquad\qquad\qquad\qquad\quad \text{on } \partial\Omega,\ t \in [0,T].
\end{cases}
\tag{2.22}
$$

Therefore, the solutions of system (2.22) can be written as

$$
\begin{aligned}
\tilde{V}_1 =& H_{\alpha_1}(z)z^{\alpha_1-1}(-az^{-\alpha_1}G_1(0) - z^{-\alpha_1}AG_1(0)) \\
& + H_{\alpha_1}(z)z^{\alpha_1-1}(az^{-\alpha_2}G_2(0)) \\
& + aH(z)z^{\alpha_1-1}(-az^{-\alpha_2}G_2(0) - z^{-\alpha_2}AG_2(0)) \\
& + aH(z)z^{\alpha_1-1}(az^{-\alpha_1}G_1(0)) \\
=& \tilde{V}_{1,1} + \tilde{V}_{1,2} + \tilde{V}_{1,3} + \tilde{V}_{1,4}, \\
\tilde{V}_2 =& aH(z)z^{\alpha_2-1}(-az^{-\alpha_1}G_1(0) - z^{-\alpha_1}AG_1(0)) \\
& + aH(z)z^{\alpha_2-1}(az^{-\alpha_2}G_2(0)) \\
& + H_{\alpha_2}(z)z^{\alpha_2-1}(-az^{-\alpha_2}G_2(0) - z^{-\alpha_2}AG_2(0)) \\
& + H_{\alpha_2}(z)z^{\alpha_2-1}(az^{-\alpha_1}G_1(0)) \\
=& \tilde{V}_{2,1} + \tilde{V}_{2,2} + \tilde{V}_{2,3} + \tilde{V}_{2,4}.
\end{aligned}
\tag{2.23}
$$

Let $\bar{V}_1^{-1} = \bar{V}_2^{-1} = 0$. Then (2.21) can be rewritten as

$$
\begin{cases}
\dfrac{1}{\tau}\left(\dfrac{3}{2}\bar{V}_1^n - 2\bar{V}_1^{n-1} + \dfrac{1}{2}\bar{V}_1^{n-2}\right) + a\sum_{i=0}^{n-1} d_{2,i}^{1-\alpha_1}\bar{V}_1^{n-i} \\[2mm]
\quad + \sum_{i=0}^{n-1} d_{2,i}^{1-\alpha_1} A\bar{V}_1^{n-i} - a\sum_{i=0}^{n-1} d_{2,i}^{1-\alpha_2}\bar{V}_2^{n-i} \\[2mm]
\quad = -(a+A)\left(\sum_{i=0}^{n-1} d_{2,i}^{1-\alpha_1} + \dfrac{1}{2}d_{2,n-1}^{1-\alpha_1}\right)G_1(0) \\[2mm]
\qquad\qquad + a\left(\sum_{i=0}^{n-1} d_{2,i}^{1-\alpha_2} + \dfrac{1}{2}d_{2,n-1}^{1-\alpha_2}\right)G_2(0) \qquad\quad \text{in } \Omega,\; n \geq 1, \\[4mm]
\dfrac{1}{\tau}\left(\dfrac{3}{2}\bar{V}_2^n - 2\bar{V}_2^{n-1} + \dfrac{1}{2}\bar{V}_2^{n-2}\right) + a\sum_{i=0}^{n-1} d_{2,i}^{1-\alpha_2}\bar{V}_2^{n-i} \\[2mm]
\quad + \sum_{i=0}^{n-1} d_{2,i}^{1-\alpha_2} A\bar{V}_2^{n-i} - a\sum_{i=0}^{n-1} d_{2,i}^{1-\alpha_1}\bar{V}_1^{n-i} \\[2mm]
\quad = -(a+A)\left(\sum_{i=0}^{n-1} d_{2,i}^{1-\alpha_2} + \dfrac{1}{2}d_{2,n-1}^{1-\alpha_2}\right)G_2(0) \\[2mm]
\qquad\qquad + a\left(\sum_{i=0}^{n-1} d_{2,i}^{1-\alpha_1} + \dfrac{1}{2}d_{2,n-1}^{1-\alpha_1}\right)G_1(0) \qquad\quad \text{in } \Omega,\; n \geq 1, \\[4mm]
\bar{V}_1^0 = \bar{V}_2^0 = 0 \qquad\qquad\qquad\qquad\qquad\qquad\quad \text{in } \Omega, \\[2mm]
\bar{V}_1^n = \bar{V}_2^n = 0 \qquad\qquad\qquad\qquad\qquad\qquad\quad \text{on } \partial\Omega,\; n \geq 0.
\end{cases}
$$

$$(2.24)$$

Similar to the derivation of (2.14) and applying

$$
\zeta\left(\dfrac{3}{2} + \sum_{n=1}^{\infty} \zeta^n\right) = \left(\dfrac{3-\zeta}{2(1-\zeta)}\right)\zeta =: \nu(\zeta)
$$

and $\delta_{\tau,2}(\zeta) = \frac{(1-\zeta)+(1-\zeta)^2/2}{\tau}$, we can get

$$\sum_{n=1}^{\infty} \bar{V}_1^n \zeta^n = H_{\alpha_1}(\delta_{\tau,2}(\zeta))(\delta_{\tau,2}(\zeta))^{\alpha_1-1}$$

$$\cdot (-(a+A)(\delta_{\tau,2}(\zeta))^{1-\alpha_1} \nu(\zeta) G_1(0))$$
$$+ H_{\alpha_1}(\delta_{\tau,2}(\zeta))(\delta_{\tau,2}(\zeta))^{\alpha_1-1} a(\delta_{\tau,2}(\zeta))^{1-\alpha_2} \nu(\zeta) G_2(0)$$
$$+ aH(\delta_{\tau,2}(\zeta))(\delta_{\tau,2}(\zeta))^{\alpha_1-1}(-(a+A)(\delta_{\tau,2}(\zeta))^{1-\alpha_2} \nu(\zeta) G_2(0))$$
$$+ aH(\delta_{\tau,2}(\zeta))(\delta_{\tau,2}(\zeta))^{\alpha_1-1} a(\delta_{\tau,2}(\zeta))^{1-\alpha_1} \nu(\zeta) G_1(0)$$
$$= \sum_{n=1}^{\infty} \bar{V}_{1,1}^n \zeta^n + \sum_{n=1}^{\infty} \bar{V}_{1,2}^n \zeta^n + \sum_{n=1}^{\infty} \bar{V}_{1,3}^n \zeta^n + \sum_{n=1}^{\infty} \bar{V}_{1,4}^n \zeta^n,$$

$$\sum_{n=1}^{\infty} \bar{V}_2^n \zeta^n = aH(\delta_{\tau,2}(\zeta))(\delta_{\tau,2}(\zeta))^{\alpha_2-1}$$

$$\cdot (-(a+A)(\delta_{\tau,2}(\zeta))^{1-\alpha_1} \nu(\zeta) G_1(0))$$
$$+ aH(\delta_{\tau,2}(\zeta))(\delta_{\tau,2}(\zeta))^{\alpha_2-1} a(\delta_{\tau,2}(\zeta))^{1-\alpha_2} \nu(\zeta) G_2(0)$$
$$+ H_{\alpha_2}(\delta_{\tau,2}(\zeta))(\delta_{\tau,2}(\zeta))^{\alpha_2-1}(-(a+A)(\delta_{\tau,2}(\zeta))^{1-\alpha_2} \nu(\zeta) G_2(0))$$
$$+ H_{\alpha_2}(\delta_{\tau,2}(\zeta))(\delta_{\tau,2}(\zeta))^{\alpha_2-1} a(\delta_{\tau,2}(\zeta))^{1-\alpha_1} \nu(\zeta) G_1(0)$$
$$= \sum_{n=1}^{\infty} V_{2,1}^n \zeta^n + \sum_{n=1}^{\infty} \bar{V}_{2,2}^n \zeta^n + \sum_{n=1}^{\infty} \bar{V}_{2,3}^n \zeta^n + \sum_{n=1}^{\infty} \bar{V}_{2,4}^n \zeta^n.$$

$$(2.25)$$

To get error estimate between $V_1(t_n)$ and $\bar{V}_1^n$, we need to get the error estimates between $V_{1,i}(t_n)$ and $\bar{V}_{1,i}^n$ ($i = 1, 2, 3, 4$). Using Eq. (2.25) and denoting $\mu(\zeta)$ as $\mu(\zeta) = \tau \delta_{\tau,2}(\zeta) \nu(\zeta) = \frac{\zeta(3-\zeta)^2}{4}$, when $\xi_\tau = e^{-(\kappa+1)\tau}$, one has

$$\bar{V}_{1,1}^n = \frac{1}{2\pi \mathbf{i}} \int_{|\zeta|=\xi_\tau} \zeta^{-n-1} H_{\alpha_1}(\delta_{\tau,2}(\zeta))(\delta_{\tau,2}(\zeta))^{\alpha_1-1}$$
$$\cdot (-(a+A))(\delta_{\tau,2}(\zeta))^{-\alpha_1} \mu(\zeta) d\zeta G_1(0).$$

By taking $\zeta = e^{-z\tau}$, using the definition of Γ^τ and deforming the contour Γ^τ to $\Gamma^\tau_{\theta,\kappa}$, we have

$$\bar{V}_{1,1}^n = \frac{1}{2\pi \mathbf{i}} \int_{\Gamma^\tau_{\theta,\kappa}} e^{zt_n} H_{\alpha_1}(\delta_{\tau,2}(e^{-z\tau}))(\delta_{\tau,2}(e^{-z\tau}))^{\alpha_1-1}$$
$$\cdot (-(a+A))(\delta_{\tau,2}(e^{-z\tau}))^{-\alpha_1} \mu(e^{-z\tau}) dz G_1(0).$$

$$(2.26)$$

Taking the inverse Laplace transform for (2.23), and combining (2.26), we obtain

$$V_{1,1}(t_n) - \bar{V}_{1,1}^n$$

$$= \frac{1}{2\pi i} \int_{\Gamma_{\theta,\kappa} \backslash \Gamma_{\theta,\kappa}^\tau} e^{zt_n} H_{\alpha_1}(z) z^{-1}(-(a+A)) dz G_1(0)$$

$$+ \frac{1}{2\pi i} \int_{\Gamma_{\theta,\kappa}^\tau} e^{zt_n} \Big(H_{\alpha_1}(z) z^{-1}(-(a+A))$$

$$- H_{\alpha_1}(\delta_{\tau,2}(e^{-z\tau}))(\delta_{\tau,2}(e^{-z\tau}))^{-1}(-(a+A)) \Big) dz G_1(0)$$

$$+ \frac{1}{2\pi i} \int_{\Gamma_{\theta,\kappa}^\tau} e^{zt_n} H_{\alpha_1}(\delta_{\tau,2}(e^{-z\tau}))(\delta_{\tau,2}(e^{-z\tau}))^{-1}(-(a+A))$$

$$\cdot (1 - \mu(e^{-z\tau})) dz G_1(0)$$

$$= I + II + III.$$

Lemma 2.3 gives

$$\|I\|_{L^2(\Omega)} \leq C \int_{\Gamma_{\theta,\kappa} \backslash \Gamma_{\theta,\kappa}^\tau} e^{-C|z|t_n} |z|^{-1} |dz| \|G_1(0)\|_{L^2(\Omega)}$$

$$\leq C\tau^2 t_n^{-2} \|G_1(0)\|_{L^2(\Omega)}.$$

The fact $\left\| \frac{d}{dz} H_{\alpha_1}(z) z^{-1}(-(a+A)) \right\| \leq C|z|^{-2}$, $\delta_{\tau,2}(e^{-z\tau}) = z + \mathcal{O}(\tau^2 z^3)$ and the mean value theorem imply

$$\|II\|_{L^2(\Omega)} \leq C\tau^2 \int_{\Gamma_{\theta,\kappa}^\tau} e^{Re(z)t_n} |z| |dz| \|G_1(0)\|_{L^2(\Omega)}$$

$$\leq C\tau^2 t_n^{-2} \|G_1(0)\|_{L^2(\Omega)}.$$

Combining the fact $\mu(e^{-z\tau}) = 1 + \mathcal{O}(z^2 \tau^2)$ [90, 124] and Lemma 2.3 yields

$$\|III\|_{L^2(\Omega)} \leq C\tau^2 \int_{\Gamma_{\theta,\kappa}^\tau} e^{Re(z)t_n} |z| |dz| \|G_1(0)\|_{L^2(\Omega)}$$

$$\leq C\tau^2 t_n^{-2} \|G_1(0)\|_{L^2(\Omega)}.$$

Thus

$$\|V_{1,1}(t_n) - \bar{V}_{1,1}^n\|_{L^2(\Omega)} \leq C\tau^2 t_n^{-2} \|G_1(0)\|_{L^2(\Omega)}.$$

Similarly

$$\|V_{1,2}(t_n) - \bar{V}_{1,2}^n\|_{L^2(\Omega)}, \|V_{1,3}(t_n) - \bar{V}_{1,3}^n\|_{L^2(\Omega)}$$

$$\leq C\tau^2 t_n^{\alpha_2-2} \|G_2(0)\|_{L^2(\Omega)},$$

$$\|V_{1,4}(t_n) - \bar{V}_{1,4}^n\|_{L^2(\Omega)} \leq C\tau^2 t_n^{\alpha_1+\alpha_2-2} \|G_1(0)\|_{L^2(\Omega)}.$$

By the fact $T/t > 1$, we have

$$\|V_1(t_n) - \bar{V}_1^n\|_{L^2(\Omega)} \le C\tau^2 \left(t_n^{-2}\|G_1(0)\|_{L^2(\Omega)} + t_n^{\alpha_2-2}\|G_2(0)\|_{L^2(\Omega)} \right).$$

The following estimate can be got similarly

$$\|V_2(t_n) - \bar{V}_2^n\|_{L^2(\Omega)} \le C\tau^2 \left(t_n^{\alpha_1-2}\|G_1(0)\|_{L^2(\Omega)} + t_n^{-2}\|G_2(0)\|_{L^2(\Omega)} \right).$$

Thus, the proof is completed. □

2.1.5 Numerical Experiments

In this section, we verify the effectiveness of the numerical schemes by performing the one- and two-dimensional numerical experiments. Here, we use finite element method to discretize Laplace operator in (2.5). To get the temporal errors, we set

$$E_{1,\tau} = \|G_{1,\tau} - G_{1,\tau/2}\|_{L^2(\Omega)}, \quad E_{2,\tau} = \|G_{2,\tau} - G_{2,\tau/2}\|_{L^2(\Omega)},$$

where $G_{1,\tau}$ and $G_{2,\tau}$ are the numerical solutions of G_1 and G_2 at the fixed time t with step size τ. The temporal convergence rates can be, respectively, calculated by

$$\text{Rate} = \frac{\ln(E_{i,\tau}/E_{i,\tau/2})}{\ln(2)}, \quad i = 1, 2.$$

2.1.5.1 One-dimensional cases

Example 2.1 *Here, we solve the system (2.5) by the SBD scheme with $a = -2$. Use the numerical solutions with $\tau = 1/640$ and $h = 1/1024$ as the "exact" solutions. Tables 2.1 and 2.2 provide the L_2 errors and convergence rates for solving the system (2.5) for different α_1 and α_2, respectively, with initial value*

> *1. $G_1(x,0) = x(1-x)$, $G_2(x,0) = \sin(\pi x)$;*

> *2. $G_1(x,0) = (1-x)\chi_{(1/2,1)}$, $G_2(x,0) = x\chi_{(0,1/2)}$.*

The results shown in Tables 2.1 and 2.2 validate Theorem 2.3.

2.1.5.2 Two-dimensional cases

In the following examples, we use first-order scheme (2.12) to solve the system (2.1).

Example 2.2 *Consider the two-dimensional homogeneous problem (2.1) with smooth initial value. Let*

$$G_{1,0}(x,y) = x(1-x)y(1-y), \quad G_{2,0}(x,y) = x^2(1-x)y(1-y)^2,$$

Table 2.1 L_2 error and convergence rates at $t = 1$ with $h = 1/1024$ using the initial values (1)

(α_1, α_2)	$1/\tau$	20	40	80	160	320
(0.3,0.4)	$E_{1,\tau}$	2.207E-06	5.276E-07	1.276E-07	3.006E-08	5.932E-09
	Rate		2.0646	2.0477	2.0859	2.3411
	$E_{2,\tau}$	4.189E-05	1.006E-05	2.441E-06	5.763E-07	1.150E-07
	Rate		2.0574	2.0433	2.0829	2.3246
(0.7,0.8)	$E_{1,\tau}$	8.061E-06	1.900E-06	4.571E-07	1.066E-07	2.048E-08
	Rate		2.0851	2.0552	2.1008	2.3799
	$E_{2,\tau}$	1.054E-04	2.501E-05	6.052E-06	1.450E-06	3.074E-07
	Rate		2.0748	2.0469	2.0609	2.2385

Table 2.2 L_2 error and convergence rates at $t = 1$ with $h = 1/1024$ using the initial values (2)

(α_1, α_2)	$1/\tau$	20	40	80	160	320
(0.2,0.4)	$E_{1,\tau}$	3.356E-06	8.084E-07	1.964E-07	4.641E-08	9.284E-09
	Rate		2.0538	2.0413	2.0813	2.3215
	$E_{2,\tau}$	7.289E-06	1.751E-06	4.247E-07	1.002E-07	2.000E-08
	Rate		2.0577	2.0435	2.0832	2.3249
(0.6,0.8)	$E_{1,\tau}$	1.095E-05	2.621E-06	6.348E-07	1.501E-07	3.022E-08
	Rate		2.0627	2.0457	2.0804	2.3124
	$E_{2,\tau}$	1.648E-05	3.912E-06	9.469E-07	2.273E-07	4.843E-08
	Rate		2.0745	2.0467	2.0586	2.2305

$\Omega = (0,1) \times (0,1)$, $T = 0.1$, $f_1(x, y, t) = f_2(x, y, t) = 0$, and $a = -2$. To get the temporal convergence rates, we set $h = 1/256$, and the results are shown in Table 2.3, which confirm Theorem 2.1.

Example 2.3 *Consider the two-dimensional homogeneous problem* (2.1) *with nonsmooth initial value. Let*

$$G_{1,0}(x, y) = \chi_{(1/2,1) \times (0,3/4)}(x, y), \quad G_{2,0}(x, y) = \chi_{(0,3/4) \times (1/2,1)}(x, y),$$

$\Omega = (0,1) \times (0,1)$, $T = 0.1$, $f_1(x, y, t) = f_2(x, y, t) = 0$, and $a = 1$. To get the temporal convergence rates, we set $h = 1/256$ and Table 2.4 shows the results, which agree with Theorem 2.1.

Example 2.4 *Consider the inhomogeneous problem* (2.1) *with vanishing initial data. For checking the temporal convergence rates,*

Table 2.3 L_2 errors and convergence rates at t=0.1

(α_1,α_2)	$0.1/\tau$	20	40	80	160	320
(0.1,0.2)	$E_{1,\tau}$	5.400E-06	2.656E-06	1.317E-06	6.560E-07	3.273E-07
	Rate		1.0236	1.0118	1.0059	1.0030
	$E_{2,\tau}$	2.177E-06	1.066E-06	5.276E-07	2.624E-07	1.309E-07
	Rate		1.0301	1.0150	1.0075	1.0037
(0.4,0.6)	$E_{1,\tau}$	4.007E-05	1.946E-05	9.590E-06	4.760E-06	2.372E-06
	Rate		1.0420	1.0209	1.0104	1.0052
	$E_{2,\tau}$	1.508E-05	7.211E-06	3.527E-06	1.744E-06	8.671E-07
	Rate		1.0639	1.0320	1.0160	1.0080
(0.8,0.9)	$E_{1,\tau}$	2.197E-04	1.069E-04	5.260E-05	2.608E-05	1.298E-05
	Rate		1.0396	1.0226	1.0121	1.0062
	$E_{2,\tau}$	6.261E-05	2.999E-05	1.463E-05	7.221E-06	3.586E-06
	Rate		1.0618	1.0354	1.0189	1.0098

Table 2.4 L_2 errors and convergence rates at t=0.1

(α_1,α_2)	$0.1/\tau$	20	40	80	160	320
(0.1,0.2)	$E_{1,\tau}$	5.672E-05	2.790E-05	1.384E-05	6.889E-06	3.438E-06
	Rate		1.0236	1.0119	1.0059	1.0030
	$E_{2,\tau}$	1.426E-04	6.982E-05	3.455E-05	1.719E-05	8.571E-06
	Rate		1.0299	1.0149	1.0075	1.0037
(0.4,0.6)	$E_{1,\tau}$	4.243E-04	2.060E-04	1.015E-04	5.038E-05	2.510E-05
	Rate		1.0426	1.0212	1.0106	1.0053
	$E_{2,\tau}$	9.640E-04	4.619E-04	2.261E-04	1.118E-04	5.563E-05
	Rate		1.0614	1.0308	1.0154	1.0077
(0.8,0.9)	$E_{1,\tau}$	2.229E-03	1.072E-03	5.248E-04	2.596E-04	1.291E-04
	Rate		1.0562	1.0302	1.0156	1.0080
	$E_{2,\tau}$	3.497E-03	1.718E-03	8.507E-04	4.231E-04	2.110E-04
	Rate		1.0252	1.0141	1.0075	1.0039

to eliminate the influence from spatial discretization, we take $h = 1/256$ *and set*

$$f_1(x,y,t) = 10t^{0.2}\chi_{(0,1/2)\times(1/4,1)}(x,y), \quad G_{1,0}(x,y) = 0,$$
$$f_2(x,y,t) = 10t^{0.3}\chi_{(1/2,1)\times(0,1/4)}(x,y), \quad G_{2,0}(x,y) = 0,$$

$a = 0.5$, $T = 0.1$ *and* $\Omega = (0,1) \times (0,1)$. *The temporal convergence rates are provided in Table 2.5, which agree with Theorem 2.2.*

Table 2.5 L_2 errors and convergence rates at $t = 0.1$

(α_1, α_2)	$0.1/\tau$	80	160	320	640	1280
(0.1,0.2)	$E_{1,\tau}$	3.355E-05	1.876E-05	1.025E-05	5.507E-06	2.920E-06
	Rate		0.8387	0.8719	0.8966	0.9155
	$E_{2,\tau}$	5.092E-06	2.950E-06	1.642E-06	8.897E-07	4.731E-07
	Rate		0.7878	0.8452	0.8839	0.9111
(0.4,0.6)	$E_{1,\tau}$	3.147E-05	1.290E-05	5.202E-06	2.053E-06	7.877E-07
	Rate		1.2863	1.3106	1.3410	1.3821
	$E_{2,\tau}$	2.300E-05	1.092E-05	5.217E-06	2.510E-06	1.214E-06
	Rate		1.0753	1.0652	1.0559	1.0475
(0.8,0.9)	$E_{1,\tau}$	2.791E-04	1.358E-04	6.627E-05	3.244E-05	1.592E-05
	Rate		1.0395	1.0349	1.0308	1.0272
	$E_{2,\tau}$	8.463E-05	4.176E-05	2.064E-05	1.022E-05	5.066E-06
	Rate		1.0192	1.0167	1.0143	1.0121

2.2 NUMERICAL METHODS FOR THE SPACE-TIME FRACTIONAL FOKKER-PLANCK SYSTEM WITH TWO INTERNAL STATES

In (2.1), if the jump length distribution of the particles is power law instead of Gaussian, then the space derivative should be replaced with fractional Laplacian. In this section, we solve the following space-time fractional Fokker-Planck system with two internal states [189] numerically and the appropriate boundary condition is specified [49], i.e.,

$$
\begin{cases}
\mathbf{M}^T \dfrac{\partial}{\partial t}\mathbf{G} = (\mathbf{M}^T - \mathbf{I})\mathrm{diag}(\,_0D_t^{1-\alpha_1},\ _0D_t^{1-\alpha_2})\mathbf{G} \\
\qquad + \mathbf{M}^T\mathrm{diag}(-\,_0D_t^{1-\alpha_1}(-\Delta)^{s_1}, \\
\qquad -\,_0D_t^{1-\alpha_2}(-\Delta)^{s_2})\mathbf{G} & \text{in } \Omega,\ t \in (0,T], \\
\mathbf{G}(\cdot,0) = \mathbf{G}_0 & \text{in } \Omega, \\
\mathbf{G} = 0 & \text{in } \Omega^c,\ t \in [0,T],
\end{cases}
$$

$$\tag{2.27}$$

where Ω denotes a bounded domain with smooth boundary in $\mathbb{R}^n$ ($n = 1, 2, 3$); Ω^c means the complementary set of Ω in $\mathbb{R}^n$; T is a fixed final time; $\mathbf{M}$ is the transition matrix of a Markov chain, being a 2×2 invertible matrix, which can be written as

$$
\mathbf{M} = \begin{bmatrix} m & 1-m \\ 1-m & m \end{bmatrix}
$$

with $m \in [0, \frac{1}{2}) \cup (\frac{1}{2}, 1]$; $\mathbf{M}^T$ means the transpose of $\mathbf{M}$; $\mathbf{G} = [G_1, G_2]^T$ denotes the solution of the system (2.27); $\mathbf{G_0} = [G_{1,0}, G_{2,0}]^T$ is the initial value; $\mathbf{I}$ is an identity matrix; "diag" denotes a diagonal matrix formed from its vector argument; $_0D_t^{1-\alpha_i}$ $(i = 1, 2)$ are the Riemann-Liouville fractional derivatives defined by [150]

$$_0D_t^{1-\alpha_i}G = \frac{1}{\Gamma(\alpha_i)} \frac{\partial}{\partial t} \int_0^t (t - \xi)^{\alpha_i - 1} G(\xi) d\xi, \qquad \alpha_i \in (0, 1); \quad (2.28)$$

and $(-\Delta)^{s_i}$ $(i = 1, 2)$ are the fractional Laplacians given as

$$(-\Delta)^{s_i} u(x) = c_{n,s_i} \text{P.V.} \int_{\mathbb{R}^n} \frac{u(x) - u(y)}{|x - y|^{n+2s_i}} dy, \qquad s_i \in (0, 1),$$

where $c_{n,s_i} = \frac{2^{2s_i} s_i \Gamma(n/2+s_i)}{\pi^{n/2}\Gamma(1-s_i)}$ and P.V. denotes the principal value integral. Without loss of generality, we set $s_1 \leq s_2$.

2.2.1 Regularity of the Solution

In this section, we mainly provide the regularity of the system (2.27).

2.2.1.1 Preliminaries

In the following, we abbreviate $G_1(\cdot, t)$ and $G_2(\cdot, t)$ to $G_1(t)$ and $G_2(t)$, respectively. For convenience, we denote by A_i the fractional Laplacian $(-\Delta)^{s_i}$ $(i = 1, 2)$ with homogeneous Dirichlet boundary condition.

Then we recall some fractional Sobolev spaces [3–6, 51, 117]. Let $\Omega \subset \mathbb{R}^n$ $(n = 1, 2, 3)$ be an open set and $s \in (0, 1)$. Then we can define the fractional Sobolev space $H^s(\Omega)$

$$H^s(\Omega) = \left\{ w \in L^2(\Omega) : \right.$$

$$|w|_{H^s(\Omega)} = \left(\int \int_{\Omega^2} \frac{|w(x) - w(y)|^2}{|x - y|^{n+2s}} dx dy \right)^{1/2} < \infty \left. \right\},$$

whose norm can be written as $\| \cdot \|_{H^s(\Omega)} = \| \cdot \|_{L^2(\Omega)} + | \cdot |_{H^s(\Omega)}$. As for $s > 1$ and $s \notin \mathbb{N}$, the $H^s(\Omega)$ can be defined as

$$H^s(\Omega) = \{w \in H^{\lfloor s \rfloor}(\Omega) : |D^\alpha w|_{H^\sigma(\Omega)} < \infty \text{ for all } \alpha \text{ s.t. } |\alpha| = \lfloor s \rfloor\},$$

where $\sigma = s - \lfloor s \rfloor$ and $\lfloor s \rfloor$ means the biggest integer not larger than s. For the functions in $H^s(\mathbb{R}^n)$ with support in $\bar{\Omega}$, we can define $\hat{H}^s(\Omega)$ as

$$\hat{H}^s(\Omega) = \{w \in H^s(\mathbb{R}^n) : \mathbf{supp}\ w \subset \bar{\Omega}\},$$

with inner product

$$\langle u, w \rangle_s := \frac{c_{n,s}}{2}$$
$$\int \int_{(\mathbb{R}^n \times \mathbb{R}^n) \backslash (\Omega^c \times \Omega^c)} \frac{(u(x) - u(y))(w(x) - w(y))}{|x - y|^{n+2s}} dy dx.$$
$$(2.29)$$

Remark 2.2 *From [4], the norm induced by (2.29) is a multiple of the $H^s(\mathbb{R}^n)$-seminorm, which is equivalent to the full $H^s(\mathbb{R}^n)$-norm on this space. Moreover, according to [117], $\hat{H}^s(\Omega) = H^s(\Omega)$ for $s \in (0, 1/2)$.*

Next we recall some properties of the fractional Laplacian. From Reference [4], one has $(-\Delta)^s : H^l(\mathbb{R}^n) \to H^{l-2s}(\mathbb{R}^n)$ is a bounded and invertible operator. Besides, the solution of the following problem

$$\begin{cases} (-\Delta)^s u = g & \text{in } \Omega, \\ u = 0 & \text{in } \Omega^c, \end{cases} \qquad (2.30)$$

satisfies following estimates.

Theorem 2.4 ([75]) *Let $\Omega \subset \mathbb{R}^n$ be a bounded domain with smooth boundary, $g \in H^\sigma(\Omega)$ for some $\sigma \geq -s$ and consider $u \in \hat{H}^s(\Omega)$ as the solution of the Dirichlet problem (2.30). Then, there exists a constant C such that*

$$|u|_{H^{s+\gamma}(\mathbb{R}^n)} \leq C\|g\|_{H^\sigma(\Omega)},$$

where $\gamma = \min(s + \sigma, 1/2 - \epsilon)$ with $\epsilon > 0$ arbitrarily small.

2.2.1.2   A priori estimate of the solution

By a similar argument for (2.5), the system (2.27) can be rewritten as

$$
\begin{cases}
\dfrac{\partial G_1}{\partial t} + a\,_0D_t^{1-\alpha_1}G_1 + \,_0D_t^{1-\alpha_1}A_1G_1 \\
\qquad = a\,_0D_t^{1-\alpha_2}G_2 & \text{in } \Omega,\; t \in (0,T], \\[2mm]
\dfrac{\partial G_2}{\partial t} + a\,_0D_t^{1-\alpha_2}G_2 + \,_0D_t^{1-\alpha_2}A_2G_2 \\
\qquad = a\,_0D_t^{1-\alpha_1}G_1 & \text{in } \Omega,\; t \in (0,T], \\[2mm]
\mathbf{G}(\cdot,0) = \mathbf{G}_0 & \text{in } \Omega, \\[2mm]
\mathbf{G} = 0 & \text{in } \Omega^c,\; t \in [0,T],
\end{cases}
\tag{2.31}
$$

where $a = \frac{1-m}{2m-1}$, $m \in [0,1/2) \cup (1/2, 1]$, and A_i mean the fractional Laplacians $(-\Delta)^{s_i}$ $(i = 1, 2)$ with homogeneous Dirichlet boundary condition.

Similar to the derivation of (2.7), we have

$$
\begin{aligned}
\tilde{G}_1 &= H(z, A_1, \alpha_1, \alpha_1 - 1)G_{1,0} + aH(z, A_1, \alpha_1, \alpha_1 - \alpha_2)\tilde{G}_2, \\
\tilde{G}_2 &= H(z, A_2, \alpha_2, \alpha_2 - 1)G_{2,0} + aH(z, A_2, \alpha_2, \alpha_2 - \alpha_1)\tilde{G}_1,
\end{aligned}
\tag{2.32}
$$

which can also be written as

$$
\begin{aligned}
\tilde{G}_1 &= H(z, A_1, \alpha_1, \alpha_1 - 1)G_{1,0} + aH(z, A_1, \alpha_1, \alpha_1 - \alpha_2) \\
&\quad \cdot (H(z, A_2, \alpha_2, \alpha_2 - 1)G_{2,0} \\
&\qquad + aH(z, A_2, \alpha_2, \alpha_2 - \alpha_1)\tilde{G}_1), \\
\tilde{G}_2 &= H(z, A_2, \alpha_2, \alpha_2 - 1)G_{2,0} + aH(z, A_2, \alpha_2, \alpha_2 - \alpha_1) \\
&\quad \cdot (H(z, A_1, \alpha_1, \alpha_1 - 1)G_{1,0} \\
&\qquad + aH(z, A_1, \alpha_1, \alpha_1 - \alpha_2)\tilde{G}_2).
\end{aligned}
\tag{2.33}
$$

Here

$$
H(z, A, \alpha, \beta) = z^\beta \left(z^\alpha + a + A\right)^{-1}.
\tag{2.34}
$$

Next we define the operator $A = (-\Delta)^s$ with homogeneous Dirichlet boundary condition and $s \in (0, 1)$, which satisfies following resolvent estimate (see Appendix B in [4] or [124])

$$
\|(z^\alpha + A)^{-1}\|_{L^2(\Omega) \to L^2(\Omega)} \leq C|z|^{-\alpha} \quad \forall z \in \Sigma_\theta, \quad \alpha \in (0, 1)
\tag{2.35}
$$

with
$$\Sigma_\theta = \{z \in \mathbb{C} : z \neq 0, |\arg z| \leq \theta\}.$$

Now some estimates in different norms for $H(z, A, \alpha, \beta)$ ($\alpha \in (0,1)$, $\beta > -1$) are provided.

Lemma 2.4 *Let A be the fractional Laplacian $(-\Delta)^s$ with homogeneous Dirichlet boundary condition and $s \in (0,1)$. When $z \in \Sigma_{\theta,\kappa}$, $\pi/2 < \theta < \pi$, κ is large enough, $\alpha \in (0,1)$ and $\beta > -1$, the following estimates hold*

$$\|H(z, A, \alpha, \beta)\|_{L^2(\Omega) \to L^2(\Omega)} \leq C|z|^{\beta-\alpha},$$
$$\|AH(z, A, \alpha, \beta)\|_{L^2(\Omega) \to L^2(\Omega)} \leq C|z|^\beta,$$

where $H(z, A, \alpha, \beta)$ is defined in (2.34).

Proof: Let $u = H(z, A, \alpha, \beta)v$. After doing simple calculations, one gets
$$u = (z^\alpha + A)^{-1}(-au + z^\beta v).$$

By using the resolvent estimate (2.35), the $\|u\|_{L^2(\Omega)}$ satisfies

$$\|u\|_{L^2(\Omega)} \leq C|z|^{-\alpha}(|a|\|u\|_{L^2(\Omega)} + |z|^\beta\|v\|_{L^2(\Omega)}).$$

Taking κ large enough and $|z| > \kappa$, the first desired estimate can be obtained. By the fact $AH(z, A, \alpha, \beta) = z^\beta(I - (z^\alpha + a)H(z, A, \alpha, 0))$, where I denotes the identity operator, the second estimate can be easily obtained. □

Lemma 2.5 *Let A be the fractional Laplacian $(-\Delta)^s$ with homogeneous Dirichlet boundary condition, $s \in (0, 1/2)$, $\alpha \in (0,1)$, and $\beta > -1$. When $z \in \Sigma_{\theta,\kappa}$, $\pi/2 < \theta < \pi$ and κ is large enough, one has*

$$\|H(z, A, \alpha, \beta)\|_{\hat{H}^\sigma(\Omega) \to \hat{H}^\sigma(\Omega)} \leq C|z|^{\beta-\alpha},$$
$$\|AH(z, A, \alpha, \beta)\|_{\hat{H}^\sigma(\Omega) \to \hat{H}^\sigma(\Omega)} \leq C|z|^\beta,$$

where the definition of $H(z, A, \alpha, \beta)$ can be seen in (2.34) and $\sigma \in [0, 1/2 + s)$. Furthermore, it holds

$$\|AH(z, A, \alpha, \beta)\|_{\hat{H}^{\dot\sigma+2\mu s}(\Omega) \to \hat{H}^{\dot\sigma}(\Omega)} \leq C|z|^{\beta-\mu\alpha},$$

where $\dot\sigma \in [0, 1/2)$ and $\mu \in [0,1]$.

Proof: Assume $u = H(z, A, \alpha, \beta)v$ with $v = 0$ in Ω^c. Theorem 2.4 and Lemma 2.4 give

$$\|u\|_{\hat{H}^{2s}(\Omega)} \leq C\|Au\|_{L^2(\Omega)} = C\|AH(z, A, \alpha, \beta)v\|_{L^2(\Omega)}$$
$$\leq C|z|^{\beta-\alpha}\|Av\|_{L^2(\Omega)} \leq C|z|^{\beta-\alpha}\|v\|_{\hat{H}^{2s}(\Omega)}, \tag{2.36}$$

which implies

$$\|H(z, A, \alpha, \beta)\|_{\hat{H}^{2s}(\Omega)\to\hat{H}^{2s}(\Omega)} \leq C|z|^{\beta-\alpha}.$$

Iterating above process for $m - 1$ times, we can obtain following estimates

$$\|H(z, A, \alpha, \beta)\|_{\hat{H}^{2ms}(\Omega)\to\hat{H}^{2ms}(\Omega)} \leq C|z|^{\beta-\alpha},$$

where m is the maximum positive integer such that $(2m-1)s < 1/2$. By Lemma 2.4 and the interpolation property [6], there holds

$$\|H(z, A, \alpha, \beta)\|_{\hat{H}^{\sigma}(\Omega)\to\hat{H}^{\sigma}(\Omega)} \leq C|z|^{\beta-\alpha}, \quad \sigma \in [0, 2ms]. \tag{2.37}$$

For $2ms < 1/2$ and $(2m + 1)s > 1/2$, taking $\sigma = 1/2 - 2s - \epsilon$ in (2.37), using Theorem 2.4 and Lemma 2.4 again, we obtain

$$\|u\|_{\hat{H}^{1/2-\epsilon}(\Omega)} \leq C\|Au\|_{\hat{H}^{\sigma}(\Omega)} = C\|AH(z, A, \alpha, \beta)v\|_{\hat{H}^{\sigma}(\Omega)}$$
$$\leq C|z|^{\beta-\alpha}\|Av\|_{\hat{H}^{\sigma}(\Omega)} \leq C|z|^{\beta-\alpha}\|v\|_{\hat{H}^{1/2-\epsilon}(\Omega)},$$

which implies

$$\|H(z, A, \alpha, \beta)\|_{\hat{H}^{1/2-\epsilon}(\Omega)\to\hat{H}^{1/2-\epsilon}(\Omega)} \leq C|z|^{\beta-\alpha}.$$

According to the interpolation property, one has

$$\|H(z, A, \alpha, \beta)\|_{\hat{H}^{\sigma}(\Omega)\to\hat{H}^{\sigma}(\Omega)} \leq C|z|^{\beta-\alpha}, \quad \sigma \in [0, 1/2).$$

Similarly, for $s \in (0, 1/2)$, one obtains

$$\|H(z, A, \alpha, \beta)\|_{\hat{H}^{1/2+s-\epsilon}(\Omega)\to\hat{H}^{1/2+s-\epsilon}(\Omega)} \leq C|z|^{\beta-\alpha},$$

by taking $\sigma = 1/2 - s - \epsilon$ and using (2.36). And by interpolation property, there holds

$$\|H(z, A, \alpha, \beta)\|_{\hat{H}^{\sigma}(\Omega)\to\hat{H}^{\sigma}(\Omega)} \leq C|z|^{\beta-\alpha}, \quad \sigma \in [0, 1/2 + s).$$

Noting that $AH(z, A, \alpha, \beta) = z^\beta(I - (z^\alpha + a)H(z, A, \alpha, 0))$ with I being the identity operator, the second estimate can be got similarly.

On the other hand, let $u = AH(z, A, \alpha, \beta)v$ and $v = 0$ in Ω^c. Similar to the derivation of the first two estimates, we can get the third estimation easily. □

Lemma 2.6 *Let κ satisfy the conditions needed in Lemma 2.4 and $\Omega \subset \mathbb{R}^n$. Then*

$$\int_{\Gamma_{\theta,\kappa}} |e^{zt}||z|^\alpha|dz| \leq Ct^{-\alpha-1}.$$

Proof: Doing simple calculations and taking $r = |z|$, we have

$$\int_{\Gamma_{\theta,\kappa}} |e^{zt}||z|^\alpha|dz| = \int_\kappa^\infty e^{r\cos(\theta)t} r^\alpha dr + \kappa^{1+\alpha} \int_{-\theta}^\theta e^{\kappa\cos(\eta)t} d\eta$$

$$\leq Ct^{-\alpha-1} + C\kappa^{1+\alpha}.$$

When $\alpha \geq -1$, we get the desired estimate by the fact $T/t > 1$. And when $\alpha < -1$, the desired estimate can be obtained by taking $\kappa > 1/t$. □

Next, we provide the following Grönwall inequality, which can be got by the similar arguments provided in [57].

Lemma 2.7 *Let the function $\phi(t) \geq 0$ be continuous for $0 < t \leq T$. If*

$$\phi(t) \leq \sum_{k=1}^N a_k t^{-1+\alpha_k} + b\int_0^t (t-s)^{-1+\beta}\phi(s)ds, \ 0 < t \leq T,$$

for some positive constants $\{a_k\}_{k=1}^N$, $\{\alpha_k\}_{k=1}^N$, b and β, then there is a positive constant $C = C(b, T, \{\alpha_k\}_{k=1}^N, \beta)$ such that

$$\phi(t) \leq C\sum_{k=1}^N a_k t^{-1+\alpha_k} \quad for \ \ 0 < t \leq T.$$

Now, we present the a priori estimates for the solutions G_1 and G_2 of (2.31) with nonsmooth initial value.

Theorem 2.5 *Let $\gamma_1 = \min(s_1, 1/2-\epsilon)$ and $\gamma_2 = \min(s_2, 1/2-\epsilon)$. If $G_{1,0}, \ G_{2,0} \in L^2(\Omega)$, then*

$$\|G_1(t)\|_{L^2(\Omega)} \leq C\|G_{1,0}\|_{L^2(\Omega)} + C\|G_{2,0}\|_{L^2(\Omega)},$$
$$\|G_2(t)\|_{L^2(\Omega)} \leq C\|G_{1,0}\|_{L^2(\Omega)} + C\|G_{2,0}\|_{L^2(\Omega)};$$

and

$$\|G_1(t)\|_{\hat{H}^{s_1+\gamma_1}(\Omega)} \le Ct^{-\alpha_1}\|G_{1,0}\|_{L^2(\Omega)} + Ct^{\min(0,\alpha_2-\alpha_1)}\|G_{2,0}\|_{L^2(\Omega)},$$
$$\|G_2(t)\|_{\hat{H}^{s_2+\gamma_2}(\Omega)} \le Ct^{\min(0,\alpha_1-\alpha_2)}\|G_{1,0}\|_{L^2(\Omega)} + Ct^{-\alpha_2}\|G_{2,0}\|_{L^2(\Omega)}.$$

Proof: With the help of (2.33), Lemmas 2.4, 2.6 and taking the inverse Laplace transform for (2.33), one has

$$\|G_1(t)\|_{L^2(\Omega)} \le C\|G_{1,0}\|_{L^2(\Omega)} + Ct^{\alpha_2}\|G_{2,0}\|_{L^2(\Omega)}$$
$$+ C\int_0^t (t-s)^{\alpha_1+\alpha_2-1}\|G_1(s)\|_{L^2(\Omega)}ds,$$
$$\|G_2(t)\|_{L^2(\Omega)} \le Ct^{\alpha_1}\|G_{1,0}\|_{L^2(\Omega)} + C\|G_{2,0}\|_{L^2(\Omega)}$$
$$+ C\int_0^t (t-s)^{\alpha_1+\alpha_2-1}\|G_2(s)\|_{L^2(\Omega)}ds.$$

Using Lemma 2.7 and the fact $T/t > 1$, one gets the desired L^2 estimates. Similarly, one can obtain the desired estimates by acting A_i on both sides of (2.33) respectively, using Lemmas 2.4, 2.6 and L^2 estimates. □

Lastly, we provide discussions on the regularity of the solutions when $s_1 \in (0, 1/2)$ in detail.

Theorem 2.6 *Assume $s_1 \le s_2 < 1/2$. If $G_{1,0}$, $G_{2,0} \in \hat{H}^\sigma(\Omega)$ and $\sigma \in (0, 1/2)$, then there hold*

$$\|G_1(t)\|_{\hat{H}^{s_1+\gamma_1}(\Omega)} \le Ct^{-\alpha_1}\|G_{1,0}\|_{\hat{H}^\sigma(\Omega)} + Ct^{\min(0,\alpha_2-\alpha_1)}\|G_{2,0}\|_{\hat{H}^\sigma(\Omega)},$$
$$\|G_2(t)\|_{\hat{H}^{s_2+\gamma_2}(\Omega)} \le Ct^{\min(0,\alpha_1-\alpha_2)}\|G_{1,0}\|_{\hat{H}^\sigma(\Omega)} + Ct^{-\alpha_2}\|G_{2,0}\|_{\hat{H}^\sigma(\Omega)},$$

where $\gamma_i = \min(1/2 - \epsilon, s_i + \sigma)$ $(i = 1, 2)$.

Remark 2.3 *Similarly, following the proof of Theorem 2.5, we can get Theorem 2.6.*

Theorem 2.7 *Assume $s_1 \le s_2 < 1/2$, $G_{i,0} \in \hat{H}^{\sigma_i}(\Omega)$, and $\sigma_i \in (0, 1/2)$ $(i = 1, 2)$. Denote $\mu_1 = \max(\frac{\alpha_1-\alpha_2}{\alpha_1} + \epsilon, 0)$, $\mu_2 = \max(\frac{\alpha_2-\alpha_1}{\alpha_2} + \epsilon, 0)$, and $\bar{\gamma}_i = \min(1/2 - \epsilon, s_i + \sigma_i)$ $(i = 1, 2)$. Then the following three cases hold*

- *If $\sigma_1 + 2\mu_1 s_1 < s_2 + \bar{\gamma}_2$ and $\sigma_2 + 2\mu_2 s_2 < s_1 + \bar{\gamma}_1$, then we have*

$$\|G_1(t)\|_{\hat{H}^{s_1+\gamma_1}(\Omega)} \le Ct^{-\alpha_1}\|G_{1,0}\|_{\hat{H}^{\sigma_1}(\Omega)} + Ct^{-\alpha_2}\|G_{2,0}\|_{\hat{H}^{\sigma_2}(\Omega)},$$
$$\|G_2(t)\|_{\hat{H}^{s_2+\gamma_2}(\Omega)} \le Ct^{-\alpha_1}\|G_{1,0}\|_{\hat{H}^{\sigma_1}(\Omega)} + Ct^{-\alpha_2}\|G_{2,0}\|_{\hat{H}^{\sigma_2}(\Omega)}.$$

- *Assume $\sigma_1 > \sigma_2$. If $\sigma_1 + 2\mu_1 s_1 > s_2 + \bar{\gamma}_2$ or $\sigma_2 + 2\mu_2 s_2 > s_1 + \bar{\gamma}_1$, then we get*

$$\|G_1(t)\|_{\hat{H}^{s_1+\gamma_1}(\Omega)} \leq Ct^{min(-\alpha_1,\alpha_1-\alpha_2)}\|G_{1,0}\|_{\hat{H}^{\sigma_1}(\Omega)}$$
$$+ Ct^{-\alpha_2}\|G_{2,0}\|_{\hat{H}^{\sigma_2}(\Omega)},$$
$$\|G_2(t)\|_{\hat{H}^{s_2+\bar{\gamma}_2}(\Omega)} \leq Ct^{min(0,\alpha_1-\alpha_2)}\|G_{1,0}\|_{\hat{H}^{\sigma_1}(\Omega)}$$
$$+ Ct^{-\alpha_2}\|G_{2,0}\|_{\hat{H}^{\sigma_2}(\Omega)}.$$

- *Assume $\sigma_1 < \sigma_2$. If $\sigma_1 + 2\mu_1 s_1 > s_2 + \bar{\gamma}_2$ or $\sigma_2 + 2\mu_2 s_2 > s_1 + \bar{\gamma}_1$, then we obtain*

$$\|G_1(t)\|_{\hat{H}^{s_1+\bar{\gamma}_1}(\Omega)} \leq Ct^{-\alpha_1}\|G_{1,0}\|_{\hat{H}^{\sigma_1}(\Omega)}$$
$$+ Ct^{min(0,\alpha_2-\alpha_1)}\|G_{2,0}\|_{\hat{H}^{\sigma_2}(\Omega)},$$
$$\|G_2(t)\|_{\hat{H}^{s_2+\gamma_2}(\Omega)} \leq Ct^{-\alpha_1}\|G_{1,0}\|_{\hat{H}^{\sigma_1}(\Omega)}$$
$$+ Ct^{min(-\alpha_2,\alpha_2-\alpha_1)}\|G_{2,0}\|_{\hat{H}^{\sigma_2}(\Omega)}.$$

Here $\gamma_1 = \min(1/2 - \epsilon, s_1 + \sigma_1, s_1 + s_2 + \bar{\gamma}_2 - 2\mu_1 s_1)$ and $\gamma_2 = \min(1/2 - \epsilon, s_2 + \sigma_2, s_2 + s_1 + \bar{\gamma}_1 - 2\mu_2 s_2)$.

Proof: Similar to the proof of Theorem 2.5, the estimation of the first case hold directly. As for $\sigma_1 + 2\mu_1 s_1 > s_2 + \bar{\gamma}_2$ or $\sigma_2 + 2\mu_2 s_2 > s_1 + \bar{\gamma}_1$, we consider $\sigma_1 > \sigma_2$ first. Theorem 2.6 gives

$$\|G_2(t)\|_{\hat{H}^{s_2+\bar{\gamma}_2}(\Omega)} \leq Ct^{\alpha_1-\alpha_2}\|G_{1,0}\|_{\hat{H}^{\sigma_1}(\Omega)} + Ct^{-\alpha_2}\|G_{2,0}\|_{\hat{H}^{\sigma_2}(\Omega)},$$

which implies

$$\|G_1(t)\|_{\hat{H}^{s_1+\gamma_1}(\Omega)} \leq Ct^{-\alpha_1}\|G_{1,0}\|_{\hat{H}^{\sigma_1}(\Omega)}$$
$$+ C\int_0^t (t-s)^{(\mu_1-1)\alpha_1+\alpha_2-1}\|G_2(s)\|_{\hat{H}^{s_2+\bar{\gamma}_2}(\Omega)}ds.$$

Here $\gamma_1 = \min(1/2 - \epsilon, s_1 + s_2 + \bar{\gamma}_2 - 2\mu_1 s_1, s_1 + \sigma_1)$. Similarly, we can get results for $\sigma_1 < \sigma_2$. $\square$

Theorem 2.8 *Assume $s_1 < 1/2 \leq s_2$. If $G_{i,0} \in \hat{H}^{\sigma_i}(\Omega)$, $(i = 1, 2)$, $\sigma_1 \in (0, 1/2 - s_1)$ and $\sigma_2 = 0$, then there hold*

$$\|G_1(t)\|_{\hat{H}^{s_1+\gamma_1}(\Omega)} \leq Ct^{-\alpha_1}\|G_{1,0}\|_{\hat{H}^{\sigma_1}(\Omega)} + C\|G_{2,0}\|_{\hat{H}^{\sigma_2}(\Omega)},$$
$$\|G_2(t)\|_{\hat{H}^{s_2+\gamma_2}(\Omega)} \leq Ct^{min(0,\alpha_1-\alpha_2)}\|G_{1,0}\|_{\hat{H}^{\sigma_1}(\Omega)}$$
$$+ Ct^{-\alpha_2}\|G_{2,0}\|_{\hat{H}^{\sigma_2}(\Omega)},$$

where $\gamma_i = \min(1/2 - \epsilon, s_i + \sigma_i)$ $(i = 1, 2)$.

Remark 2.4 *According to the proofs of Theorems 2.5 and 2.7, Theorem 2.8 can be proved.*

2.2.2 Space Discretization and Error Analysis

In this section, the space semidiscrete scheme of system (2.31) constructed by finite element method are provided and corresponding error estimates are discussed. Let h be the maximum diameter and $\mathcal{T}_h$ a shape regular quasi-uniform partition of the domain Ω. Denote X_h as the piecewise linear finite element space.

Furthermore, define the L^2-orthogonal projection P_h : $L^2(\Omega) \to X_h$ by

$$(P_h u, v_h) = (u, v_h) \quad \forall v_h \in X_h$$

with $(\cdot, \cdot)$ being the L^2 inner product.

The semidiscrete Galerkin scheme for system (2.31) reads: Find $G_{1,h}, \ G_{2,h} \in X_h$ satisfying

$$\begin{cases} \left(\dfrac{\partial G_{1,h}}{\partial t}, v_{1,h}\right) + a \ _0D_t^{1-\alpha_1}(G_{1,h}, v_{1,h}) + \ _0D_t^{1-\alpha_1}\langle G_{1,h}, v_{1,h}\rangle_{s_1} \\[2mm] = a \ _0D_t^{1-\alpha_2}(G_{2,h}, v_{1,h}), \\[2mm] \left(\dfrac{\partial G_{2,h}}{\partial t}, v_{2,h}\right) + a \ _0D_t^{1-\alpha_2}(G_{2,h}, v_{2,h}) + \ _0D_t^{1-\alpha_2}\langle G_{2,h}, v_{2,h}\rangle_{s_2} \\[2mm] = a \ _0D_t^{1-\alpha_1}(G_{1,h}, v_{2,h}), \end{cases}$$

$$(2.38)$$

for all $v_{1,h}, \ v_{2,h} \in X_h$. Here, we take $G_{1,h}(0) = P_h G_{1,0}$ and $G_{2,h}(0) = P_h G_{2,0}$.

Next, define the discrete operators $A_{i,h}: X_h \to X_h$ as

$$(A_{i,h}u_h, v_h) = \langle u_h, v_h\rangle_{s_i} \quad \forall u_h, \ v_h \in X_h, \ i = 1, 2.$$

Then we can rewrite (2.38) as

$$\frac{\partial G_{1,h}}{\partial t} + a \ _0D_t^{1-\alpha_1}G_{1,h} + \ _0D_t^{1-\alpha_1}A_{1,h}G_{1,h} = a \ _0D_t^{1-\alpha_2}G_{2,h},$$

$$\frac{\partial G_{2,h}}{\partial t} + a \ _0D_t^{1-\alpha_2}G_{2,h} + \ _0D_t^{1-\alpha_2}A_{2,h}G_{2,h} = a \ _0D_t^{1-\alpha_1}G_{1,h},$$

$$(2.39)$$

which leads to

$$z\tilde{G}_{1,h} + az^{1-\alpha_1}\tilde{G}_{1,h} + z^{1-\alpha_1}A_{1,h}\tilde{G}_{1,h} = az^{1-\alpha_2}\tilde{G}_{2,h} + G_{1,h}(0),$$

$$z\tilde{G}_{2,h} + az^{1-\alpha_2}\tilde{G}_{2,h} + z^{1-\alpha_2}A_{2,h}\tilde{G}_{2,h} = az^{1-\alpha_1}\tilde{G}_{1,h} + G_{2,h}(0).$$

$$(2.40)$$

In the following, to obtain the error estimate between system (2.31) and space semidiscrete scheme (2.38), we first introduce the following two lemmas.

Lemma 2.8 ([20, 64]) *For any $\phi \in \hat{H}^s(\Omega)$, $z \in \Sigma_{\theta,\kappa}$ with $\theta \in (\pi/2, \pi)$ and κ being taken to be large enough to ensure $g(z) = z^\alpha + a \in \Sigma_\theta$, where a is defined in (2.31), then we have*

$$|g(z)|\|\phi\|^2_{L^2(\Omega)} + \|\phi\|^2_{\hat{H}^s(\Omega)} \leq C \left| g(z)\|\phi\|^2_{L^2(\Omega)} + \|\phi\|^2_{\hat{H}^s(\Omega)} \right|.$$

Lemma 2.9 *Let $v \in L^2(\Omega)$, $A = (-\Delta)^s$ with homogeneous Dirichlet boundary condition, $s \in (0,1)$ and $z \in \Gamma_{\theta,\kappa}$ with $\theta \in (\pi/2, \pi)$ and κ being taken to be large enough. Denote $w = H(z, A, \alpha, 0)v$ and $w_h = H(z, A_h, \alpha, 0)P_h v$. Then one has*

$$\|w - w_h\|_{L^2(\Omega)} + h^\gamma \|w - w_h\|_{\hat{H}^s(\Omega)} \leq Ch^{2\gamma}\|v\|_{L^2(\Omega)},$$

where

$$(A_h u_h, v_h) = \langle u_h, v_h \rangle_s \quad \forall u_h, \ v_h \in X_h,$$

and $\gamma = \min(s, 1/2 - \epsilon)$ with $\epsilon > 0$ being arbitrarily small.

Remark 2.5 *The proofs of Lemmas 2.8 and 2.9 are similar to the ones in [4, 20].*

When $v \in \hat{H}^\sigma(\Omega)$, we can get following estimates similarly.

Lemma 2.10 *Let $A = (-\Delta)^s$ with homogeneous Dirichlet boundary condition, $s \in (0, 1/2)$, and $z \in \Gamma_{\theta,\kappa}$ with $\theta \in (\pi/2, \pi)$ and κ being taken to be large enough. Assume $v \in \hat{H}^\sigma(\Omega)$ with $\sigma \in (0, 1/2-s)$. Denote $w = H(z, A, \alpha, 0)v$ and $w_h = H(z, A_h, \alpha, 0)P_h v$. Then it holds*

$$\|w - w_h\|_{L^2(\Omega)} + h^s \|w - w_h\|_{\hat{H}^s(\Omega)} \leq Ch^{2s+\sigma}\|v\|_{\hat{H}^\sigma(\Omega)},$$

where

$$(A_h u_h, v_h) = \langle u_h, v_h \rangle_s \quad \forall u_h, \ v_h \in X_h.$$

Proof: Similar to the proof of Lemma 8 in Ref. [4] or Lemma 3.3 in Ref. [20], one can get the desired estimate. □

For (2.31), we provide the error estimates for the space semidiscrete scheme with nonsmooth initial values.

Theorem 2.9 *Let G_1, G_2 and $G_{1,h}$, $G_{2,h}$ be the solutions of the systems (2.31) and (2.39), respectively, $G_{1,0}$, $G_{2,0} \in L^2(\Omega)$ and $G_{1,h}(0) = P_h G_{1,0}$, $G_{2,h}(0) = P_h G_{2,0}$. Then*

$$\|G_1(t) - G_{1,h}(t)\|_{L^2(\Omega)} \leq Ch^{2\gamma_1}(t^{-\alpha_1}\|G_{1,0}\|_{L^2(\Omega)}$$
$$+ t^{\min(0,\alpha_2-\alpha_1)}\|G_{2,0}\|_{L^2(\Omega)})$$
$$+ Ch^{2\gamma_2}(\|G_{1,0}\|_{L^2(\Omega)} + \|G_{2,0}\|_{L^2(\Omega)}),$$
$$\|G_2(t) - G_{2,h}(t)\|_{L^2(\Omega)} \leq Ch^{2\gamma_1}(\|G_{1,0}\|_{L^2(\Omega)} + \|G_{2,0}\|_{L^2(\Omega)})$$
$$+ Ch^{2\gamma_2}(t^{\min(0,\alpha_1-\alpha_2)}\|G_{1,0}\|_{L^2(\Omega)} + t^{-\alpha_2}\|G_{2,0}\|_{L^2(\Omega)}),$$

where $\gamma_1 = \min(s_1, 1/2 - \epsilon)$ and $\gamma_2 = \min(s_2, 1/2 - \epsilon)$ with $\epsilon > 0$ being arbitrarily small.

Proof: According to (2.40), one gets

$$\tilde{G}_{1,h} = II(z, A_{1,h}, \alpha_1, \alpha_1 - 1)P_h G_{1,0} + aH(z, A_{1,h}, \alpha_1, \alpha_1 - \alpha_2)\tilde{G}_{2,h},$$
$$\tilde{G}_{2,h} = H(z, A_{2,h}, \alpha_2, \alpha_2 - 1)P_h G_{2,0} + aH(z, A_{2,h}, \alpha_2, \alpha_2 - \alpha_1)\tilde{G}_{1,h}.$$

Let $e_1(t) = G_1(t) - G_{1,h}(t)$ and $e_2(t) = G_2(t) - G_{2,h}(t)$. By (2.32), c_1 satisfies

$$\tilde{e}_1 = z^{\alpha_1-1}(H(z, A_1, \alpha_1, 0) - H(z, A_{1,h}, \alpha_1, 0)P_h)G_{1,0}$$
$$+ aH(z, A_1, \alpha_1, \alpha_1 - \alpha_2)\tilde{G}_2 - aH(z, A_{1,h}, \alpha_1, \alpha_1 - \alpha_2)\tilde{G}_{2,h}$$
$$= z^{\alpha_1-1}(H(z, A_1, \alpha_1, 0) - H(z, A_{1,h}, \alpha_1, 0)P_h)G_{1,0}$$
$$+ z^{\alpha_1-\alpha_2}(aH(z, A_1, \alpha_1, 0)\tilde{G}_2 - aH(z, A_{1,h}, \alpha_1, 0)P_h\tilde{G}_2)$$
$$+ aH(z, A_{1,h}, \alpha_1, \alpha_1 - \alpha_2)P_h\tilde{G}_2$$
$$- aH(z, A_{1,h}, \alpha_1, \alpha_1 - \alpha_2)\tilde{G}_{2,h}$$
$$= \sum_{i=1}^{3} \tilde{I}_i.$$

$$(2.41)$$

For I_1, Lemma 2.9 leads to

$$\|I_1\|_{L^2(\Omega)} \leq Ch^{2\gamma_1} \int_{\Gamma_{\theta,\kappa}} |e^{zt}||z|^{\alpha_1-1}|dz|\|G_{1,0}\|_{L^2(\Omega)}$$
$$\leq Ch^{2\gamma_1}t^{-\alpha_1}\|G_{1,0}\|_{L^2(\Omega)}.$$

Taking inverse Laplace transform for $\tilde{I}_2$, using Eq. (2.32), Lemma 2.9, and Theorem 2.5, one has

$$\|I_2\|_{L^2(\Omega)} \leq Ch^{2\gamma_1}(t^{-\alpha_1}\|G_{1,0}\|_{L^2(\Omega)} + t^{\min(0,\alpha_2-\alpha_1)}\|G_{2,0}\|_{L^2(\Omega)}),$$

where we have used the fact $T/t \geq 1$. As for I_3, similar to Lemma 2.4, one has

$$\|H(z, A_{i,h}, \alpha, \beta)\|_{L^2(\Omega) \to L^2(\Omega)} \leq C|z|^{\beta-\alpha}, \quad i = 1, 2.$$

Then we can get

$$\|I_3\|_{L^2(\Omega)} \leq C \int_0^t (t-s)^{\alpha_2-1} \|e_2(s)\|_{L^2(\Omega)} ds$$

by the inverse Laplace transform and the L^2 stability of projection P_h. So

$$\|e_1(t)\|_{L^2(\Omega)} \leq Ch^{2\gamma_1}(t^{-\alpha_1}\|G_{1,0}\|_{L^2(\Omega)} + t^{\min(0,\alpha_2-\alpha_1)}\|G_{2,0}\|_{L^2(\Omega)})$$
$$+ C \int_0^t (t-s)^{\alpha_2-1} \|e_2(s)\|_{L^2(\Omega)} ds. \tag{2.42}$$

Similarly, one has

$$\|e_2(t)\|_{L^2(\Omega)} \leq Ch^{2\gamma_2}(t^{-\alpha_2}\|G_{2,0}\|_{L^2(\Omega)} + t^{\min(0,\alpha_1-\alpha_2)}\|G_{1,0}\|_{L^2(\Omega)})$$
$$+ C \int_0^t (t-s)^{\alpha_1-1} \|e_1(s)\|_{L^2(\Omega)} ds. \tag{2.43}$$

Thus, combining (2.42),(2.43) and applying Lemma 2.7, the desired estimate can be obtained. $\qquad\square$

On the other hand, we have the following spatial error estimates for $s_1 < 1/2$.

Theorem 2.10 *Let G_1, G_2 and $G_{1,h}$, $G_{2,h}$ be the solutions of the systems (2.31) and (2.39), respectively. Assume $G_{i,0} \in \hat{H}^{\sigma_i}(\Omega)$ with $\sigma_i > 0$ and take $G_{i,h}(0) = P_h G_{i,0}$ $(i = 1, 2)$. Introduce $\gamma_i = \min(1/2 - \epsilon, s_i + \sigma_i)$ with $\epsilon > 0$ being arbitrarily small $(i = 1, 2)$, $\mu_1 = \max(\frac{\alpha_1-\alpha_2}{\alpha_1} + \epsilon, 0)$, $\mu_2 = \max(\frac{\alpha_2-\alpha_1}{\alpha_2} + \epsilon, 0)$, $\bar{\gamma}_1 = \min(s_1 + \sigma_1, s_1 + s_2 + \gamma_2 - 2\mu_1 s_1, 1/2 - \epsilon)$ and $\bar{\gamma}_2 = \min(s_2 + \sigma_2, s_2 + s_1 + \gamma_1 - 2\mu_2 s_2, 1/2 - \epsilon)$ with $\epsilon > 0$ being arbitrarily small. Define the following sets*

$$\mathbb{S}_1 = \{(s_1, s_2, \sigma_1, \sigma_2)|s_1 \leq s_2 < 1/2$$
$$\text{and } \sigma_1 = \sigma_2 < \max(1/2 - s_1, 1/2 - s_2)\},$$

$$\mathbb{S}_2 = \{(s_1, s_2, \sigma_1, \sigma_2) | s_1 \leq s_2 < 1/2, \ \sigma_i < 1/2 - \epsilon, \ i = 1, 2,$$
$$\sigma_1 + 2\mu_1 s_1 \leq s_2 + \gamma_2 \ \text{and} \ \sigma_2 + 2\mu_2 s_2 \leq s_1 + \gamma_1 \},$$
$$\mathbb{S}_3 = \{(s_1, s_2, \sigma_1, \sigma_2) | s_1 \leq s_2 < 1/2, \ \sigma_i < 1/2 - \epsilon,$$
$$i = 1, 2, \ \sigma_1 > \sigma_2,$$
$$\sigma_1 + 2\mu_1 s_1 > s_2 + \gamma_2 \ \text{or} \ \sigma_2 + 2\mu_2 s_2 > s_1 + \gamma_1 \},$$
$$\mathbb{S}_4 = \{(s_1, s_2, \sigma_1, \sigma_2) | s_1 \leq s_2 < 1/2, \ \sigma_i < 1/2 - \epsilon,$$
$$i = 1, 2, \ \sigma_1 < \sigma_2,$$
$$\sigma_1 + 2\mu_1 s_1 > s_2 + \gamma_2 \ \text{or} \ \sigma_2 + 2\mu_2 s_2 > s_1 + \gamma_1 \},$$
$$\mathbb{S}_5 = \{(s_1, s_2, \sigma_1, \sigma_2) | s_1 < 1/2 \leq s_2 \ \text{and} \ \sigma_1 < 1/2 - s_1, \sigma_2 = 0\}.$$

Then one has

$$\|G_1(t) - G_{1,h}(t)\|_{L^2(\Omega)} \leq Ch^{\nu_1} t^{-\alpha_1} \left(\|G_{1,0}\|_{\hat{H}^{\sigma_1}(\Omega)} + \|G_{2,0}\|_{\hat{H}^{\sigma_2}(\Omega)} \right)$$
$$+ Ch^{\nu_2} \left(\|G_{1,0}\|_{\hat{H}^{\sigma_1}(\Omega)} + \|G_{2,0}\|_{\hat{H}^{\sigma_2}(\Omega)} \right),$$
$$\|G_2(t) - G_{2,h}(t)\|_{L^2(\Omega)} \leq Ch^{\nu_1} \left(\|G_{1,0}\|_{\hat{H}^{\sigma_1}(\Omega)} + \|G_{2,0}\|_{\hat{H}^{\sigma_2}(\Omega)} \right)$$
$$+ Ch^{\nu_2} t^{-\alpha_2} \left(\|G_{1,0}\|_{\hat{H}^{\sigma_1}(\Omega)} + \|G_{2,0}\|_{\hat{H}^{\sigma_2}(\Omega)} \right),$$

where

$$\nu_1 = \begin{cases} \min(s_1 + \gamma_1, 1), & (s_1, s_2, \sigma_1, \sigma_2) \in \mathbb{S}_1 \cup \mathbb{S}_4 \cup \mathbb{S}_5, \\ s_1 + \bar{\gamma}_1, & (s_1, s_2, \sigma_1, \sigma_2) \in \mathbb{S}_2 \cup \mathbb{S}_3, \end{cases}$$

and

$$\nu_2 = \begin{cases} \min(s_2 + \gamma_2, 1), & (s_1, s_2, \sigma_1, \sigma_2) \in \mathbb{S}_1 \cup \mathbb{S}_3 \cup \mathbb{S}_5, \\ s_2 + \bar{\gamma}_2, & (s_1, s_2, \sigma_1, \sigma_2) \in \mathbb{S}_2 \cup \mathbb{S}_4. \end{cases}$$

Proof: Combining the proof of Theorem 2.9, the a priori estimate provided in Section 2.2.1, and Lemma 2.10, the desired results can be obtained. □

Remark 2.6 *For the following equation*

$$\begin{cases} \dfrac{\partial G}{\partial t} + (-\Delta)^s G = 0 & in \ \Omega \times (0, T], \\ G(0) = G_0 & in \ \Omega, \\ G = 0 & in \ \Omega^c, \end{cases}$$

the spatial semidiscrete scheme can be written as: Find $G_h \in X_h$ satisfying

$$\left(\dfrac{\partial G_h}{\partial t}, v_h \right) + {}_0 D_t^{1-\alpha} \langle G_h, v_h \rangle_s = 0,$$

for all $v_h \in X_h$. Here we take $G_h(0) = P_h G_0$. According to Lemma 2.10, if $G_0 \in \hat{H}^\sigma(\Omega)$ with $\sigma \geq 0$, the error between $G(t)$ and $G_h(t)$ can be written as

$$\|G(t) - G_h(t)\|_{L^2(\Omega)} \leq Ct^{-\alpha} h^{\min(s+\gamma, 1-\epsilon)} \|G_0\|_{\hat{H}^\sigma(\Omega)},$$

where $\gamma = \min(1/2 - \epsilon, s + \sigma)$.

2.2.3 Time Discretization and Error Analysis

In this section, we use L_1 method to discretize the Riemann-Liouville fractional derivatives and provide the error analysis for the fully discrete scheme.

Introduce the following notations as

$$
\begin{aligned}
H_1(z_1, z_2, A_1, A_2) &= ((1 + az_1 + z_1 A_1)(1 + az_2 + z_2 A_2) \\
&\quad - a^2 z_1 z_2)^{-1}, \\
H_2(z_1, z_2, A_1, A_2) &= H_1(z_1, z_2, A_1, A_2)(1 + az_1 + z_1 A_1).
\end{aligned}
\tag{2.44}
$$

Lemma 2.11 *When $z_1, z_2 \in \Sigma_{\theta,\kappa}$, $\pi/2 < \theta < \pi$ and $\kappa > |a|$, where a is defined in (2.31), there are*

$$\|H_1(z_1^{-1}, z_2^{-1}, A_1, A_2)\|_{L^2(\Omega) \to L^2(\Omega)} \leq C,$$
$$\|H_2(z_1^{-1}, z_2^{-1}, A_1, A_2)\|_{L^2(\Omega) \to L^2(\Omega)} \leq C.$$

Remark 2.7 *From Lemma 2.11, for $z \in \Sigma_{\theta,\kappa}$, $\pi/2 < \theta < \pi$ and $\kappa > \max\left(2|a|^{1/\alpha}, 2|a|^{1/\beta}\right)$, $\alpha, \beta \in (0, 1)$, there hold*

$$\|H_1(z^{-\alpha}, z^{-\beta}, A_1, A_2)\|_{L^2(\Omega) \to L^2(\Omega)} \leq C,$$
$$\|H_2(z^{-\alpha}, z^{-\beta}, A_1, A_2)\|_{L^2(\Omega) \to L^2(\Omega)} \leq C.$$

Let the time step size $\tau = T/L$, $L \in \mathbb{N}$, $t_i = i\tau$, $i = 0, 1, \ldots, L$ and $0 = t_0 < t_1 < \cdots < t_L = T$. According to [116], the Caputo fractional derivative with $\alpha \in (0, 1)$ can be approximated by L_1 scheme, i.e.,

$$
{}_0^C D_t^\alpha u(t_n) = \tau^{-\alpha} \left(b_0^{(\alpha)} u(t_n) + \sum_{j=1}^{n-1} (b_j^{(\alpha)} - b_{j-1}^{(\alpha)}) u(t_{n-j}) - b_{n-1}^{(\alpha)} u(t_0) \right)
$$
$$
+ \mathcal{O}(\tau^{2-\alpha})
$$

with

$$b_j^{(\alpha)} = ((j+1)^{1-\alpha} - j^{1-\alpha})/\Gamma(2-\alpha), \ j = 0, 1, \cdots, n-1.$$

By the following relationship

$$_0D_t^\alpha u(t) = {}_0^C D_t^\alpha u(t) + \frac{t^{-\alpha}}{\Gamma(1-\alpha)} u(0), \quad \alpha \in (0,1),$$

the Riemann-Liouville fractional derivative can be approximated by

$$_0D_t^\alpha u(t_n) = \tau^{-\alpha} \sum_{j=0}^n d_j^{(\alpha)} u(t_{n-j}) + \mathcal{O}(\tau^{2-\alpha}),$$

where

$$d_j^{(\alpha)} = \begin{cases} b_0^{(\alpha)} & \text{for } j = 0, \\ b_j^{(\alpha)} - b_{j-1}^{(\alpha)} & \text{for } 0 < j < n, \\ b_{n-1}^{(\alpha)} + \dfrac{n^{-\alpha}}{\Gamma(1-\alpha)} & \text{for } j = n. \end{cases} \tag{2.45}$$

Thus the fully discrete scheme of (2.31) can be written as

$$\begin{cases} \dfrac{G_{1,h}^n - G_{1,h}^{n-1}}{\tau} + a\tau^{\alpha_1 - 1} \displaystyle\sum_{i=0}^{n-1} d_i^{(1-\alpha_1)} G_{1,h}^{n-i} \\ \qquad + \tau^{\alpha_1 - 1} \displaystyle\sum_{i=0}^{n-1} d_i^{(1-\alpha_1)} A_{1,h} G_{1,h}^{n-i} \\ \qquad = a\tau^{\alpha_2 - 1} \displaystyle\sum_{i=0}^{n-1} d_i^{(1-\alpha_2)} G_{2,h}^{n-i}, \\ \dfrac{G_{2,h}^n - G_{2,h}^{n-1}}{\tau} + a\tau^{\alpha_2 - 1} \displaystyle\sum_{i=0}^{n-1} d_i^{(1-\alpha_2)} G_{2,h}^{n-i} \\ \qquad + \tau^{\alpha_2 - 1} \displaystyle\sum_{i=0}^{n-1} d_i^{(1-\alpha_2)} A_{2,h} G_{2,h}^{n-i} \\ \qquad = a\tau^{\alpha_1 - 1} \displaystyle\sum_{i=0}^{n-1} d_i^{(1-\alpha_1)} G_{1,h}^{n-i}, \\ G_{1,h}^0 = G_{1,h}(0), \\ G_{2,h}^0 = G_{2,h}(0), \end{cases} \tag{2.46}$$

where $G_{1,h}^n$, $G_{2,h}^n$ are the numerical solutions of G_1, G_2 at time t_n and α_1, $\alpha_2 \in (0,1)$.

Introduce $Li_p(z)$ [112] as

$$Li_p(z) = \sum_{j=1}^{\infty} \frac{z^j}{j^p},$$

which satisfies the following properties.

Lemma 2.12 ([57,87]) *For $p \neq 1,2,\cdots$, the function $Li_p(e^{-z})$ satisfies the singular expansion*

$$Li_p(e^{-z}) \sim \Gamma(1-p)z^{p-1} + \sum_{l=0}^{\infty}(-1)^l \varsigma(p-l)\frac{z^l}{l!} \qquad \text{as } z \to 0,$$

where $\varsigma(z)$ denotes the Riemann zeta function.

Lemma 2.13 ([57,87]) *Let $|z| \leq \frac{\pi}{\sin(\theta)}$ with $\theta \in (\pi/2, 5\pi/6)$ and $-1 < p < 0$. Then*

$$Li_p(e^{-z}) = \Gamma(1-p)z^{p-1} + \sum_{l=0}^{\infty}(-1)^l \varsigma(p-l)\frac{z^l}{l!}$$

converges absolutely.

Next, we provide the error estimates of the fully discrete scheme. Similar to the derivation of (2.14), the solutions of the system (2.46) can be written as

$$G_{1,h}^n = \frac{1}{2\pi i} \int_{\Gamma_{\theta,\kappa}^\tau} e^{z t_n} e^{-z\tau} \left(\frac{1 - e^{-z\tau}}{\tau} \right)^{-1}$$

$$\cdot \Bigg(H_2(\psi^{(1-\alpha_2)}(e^{-z\tau}), \psi^{(1-\alpha_1)}(e^{-z\tau}), A_{2,h}, A_{1,h}) G_{1,h}(0)$$

$$+ a H_1(\psi^{(1-\alpha_2)}(e^{-z\tau}), \psi^{(1-\alpha_1)}(e^{-z\tau}), A_{2,h}, A_{1,h})$$

$$\psi^{(1-\alpha_2)}(e^{-z\tau}) G_{2,h}(0) \Bigg) dz,$$

$$G_{2,h}^n = \frac{1}{2\pi i} \int_{\Gamma_{\theta,\kappa}^\tau} e^{z t_n} e^{-z\tau} \left(\frac{1 - e^{-z\tau}}{\tau} \right)^{-1}$$

$$\cdot \Bigg(H_1(\psi^{(1-\alpha_1)}(e^{-z\tau}), \psi^{(1-\alpha_2)}(e^{-z\tau}), A_{2,h}, A_{1,h})$$

$$\psi^{(1-\alpha_1)}(e^{-z\tau}) G_{1,h}(0)$$

$$+ a H_2(\psi^{(1-\alpha_1)}(e^{-z\tau}), \psi^{(1-\alpha_2)}(e^{-z\tau}), A_{2,h}, A_{1,h}) G_{2,h}(0) \Bigg) dz,$$

$$(2.47)$$

where

$$\psi^{(\alpha)}(\zeta) = \tau^{-\alpha} \left(\frac{1 - \zeta}{\tau} \right)^{-1} \left(\sum_{j=0}^{\infty} d_j^{(\alpha)} \zeta^j \right) \qquad \alpha \in (0, 1). \qquad (2.48)$$

As for $\psi^{(\alpha)}(\zeta)$, using the definitions of $d_j^{(\alpha)}$ and $Li_p(z)$, we have

$$\psi^{(\alpha)}(\zeta) = \tau^{-\alpha} \left(\frac{1 - \zeta}{\tau} \right)^{-1} \left(\sum_{j=1}^{\infty} (b_j^{(\alpha)} - b_{j-1}^{(\alpha)}) \zeta^j + b_0^{(\alpha)} \zeta^0 \right)$$

$$= \tau^{1-\alpha} \sum_{j=0}^{\infty} b_j^{(\alpha)} \zeta^j = \frac{\tau^{1-\alpha}}{\Gamma(2 - \alpha)} \left(\sum_{j=0}^{\infty} ((j+1)^{1-\alpha} - j^{1-\alpha}) \zeta^j \right)$$

$$= \frac{\tau^{1-\alpha}}{\Gamma(2 - \alpha)} \frac{(1 - \zeta)}{\zeta} \left(\sum_{j=0}^{\infty} j^{1-\alpha} \zeta^j \right)$$

$$= \frac{\tau^{1-\alpha}}{\Gamma(2 - \alpha)} \frac{(1 - \zeta)}{\zeta} Li_{\alpha-1}(\zeta).$$

Then we discuss the uniform lower bound of $\tau^{\alpha-1} \psi^{(\alpha)}(e^{-z\tau})$.

Lemma 2.14 ([87]) *For $z \in \Gamma_{\theta,\kappa}$ and $|z\tau| \leq \frac{\pi}{\sin(\theta)}$ with any θ close to $\pi/2$, then it holds for any $\kappa < \frac{\pi}{2\tau}$,*

$$|\tau^{\alpha-1}\psi^{(\alpha)}(e^{-z\tau})| \geq C > 0 \quad \text{and}$$

$$Re(\tau^{\alpha-1}\psi^{(\alpha)}(e^{-z\tau})) > 0 \quad \forall z \in \Gamma_{\theta,\kappa},$$

where $Re(z)$ means the real part of z.

And, the following estimate holds.

Lemma 2.15 *Let $z \in \Gamma_{\theta,\kappa}$, $|z\tau| \leq \frac{\pi}{\sin(\theta)}$ and $\theta \in (\pi/2, 5\pi/6)$. Then one has*

$$\left|\psi^{(\alpha)}(e^{-z\tau}) - z^{\alpha-1}\right| \leq C\tau|z|^{\alpha}, \quad |\psi^{(\alpha)}(e^{-z\tau})| \leq C|z|^{\alpha-1},$$

and $\psi^{(\alpha)}(e^{-z\tau}) \in \Sigma_{\theta,C\tau^{1-\alpha}}$ for any $\theta \in (\pi/2, 5\pi/6)$.

Proof: According to Lemma 2.13, there holds

$$\psi^{(\alpha)}(e^{-z\tau})$$

$$= \tau^{1-\alpha} \sum_{j=1}^{\infty} \frac{(z\tau)^j}{j!} \left[(z\tau)^{\alpha-2} + \sum_{k=0}^{\infty} \frac{(-1)^k \varsigma(-\alpha - k)}{\Gamma(2-\alpha)} \frac{(z\tau)^k}{k!} \right]$$

$$= z^{\alpha-1} + \sum_{j=2}^{\infty} \frac{z^{\alpha-2+j}\tau^{j-1}}{j!}$$

$$+ \tau^{1-\alpha} \sum_{j=1}^{\infty} \frac{(z\tau)^j}{j!} \sum_{k=0}^{\infty} \frac{(-1)^k \varsigma(-\alpha - k)}{\Gamma(2-\alpha)} \frac{(z\tau)^k}{k!}$$

$$= z^{\alpha-1} + \mathcal{O}(|z|^{\alpha}\tau).$$

On the other hand, we have

$$|\psi^{(\alpha)}(e^{-z\tau})|$$

$$\leq |z|^{\alpha-1} \left| 1 + \sum_{j=2}^{\infty} \frac{z^{j-1}\tau^{j-1}}{j!} + (z\tau)^{1-\alpha} \right.$$

$$\left. \cdot \sum_{j=1}^{\infty} \frac{(z\tau)^j}{j!} \sum_{k=0}^{\infty} \frac{(-1)^k \varsigma(-\alpha - k)}{\Gamma(2-\alpha)} \frac{(z\tau)^k}{k!} \right|$$

$$\leq C|z|^{\alpha-1}.$$

Lemma 2.14 implies

$$\left|\psi^{(\alpha)}(e^{-z\tau})\right| \geq C\tau^{1-\alpha} \quad \forall z \in \left\{ z \in \Gamma_{\theta,\kappa} : |z\tau| \leq \frac{\pi}{\sin(\theta)} \right\}$$

and $\psi^{(\alpha)}(e^{-z\tau}) \in \Sigma_{\theta,C\tau^{1-\alpha}}$ for any $\theta \in (\pi/2, 5\pi/6)$. □

Remark 2.8 *According to Lemma 2.15, one has*

$$\psi^{(1-\alpha_1)}(e^{-z\tau}) \in \Sigma_{\theta,C\tau^{\alpha_1}}, \quad \psi^{(1-\alpha_2)}(e^{-z\tau}) \in \Sigma_{\theta,C\tau^{\alpha_2}},$$

and

$$|\psi^{(1-\alpha_1)}(e^{-z\tau})| \le C|z|^{-\alpha_1}, \quad |\psi^{(1-\alpha_2)}(e^{-z\tau})| \le C|z|^{-\alpha_2}.$$

Thus we get $(\psi^{(1-\alpha_1)}(e^{-z\tau}))^{-1} \in \Sigma_{\theta,C|z|^{\alpha_1}}$ and $(\psi^{(1-\alpha_2)}(e^{-z\tau}))^{-1} \in \Sigma_{\theta,C|z|^{\alpha_2}}$.

Now we give the error estimates between the solutions of the systems (2.39) and (2.46).

Theorem 2.11 *Let $G_{1,h}$, $G_{2,h}$ and $G_{1,h}^n$, $G_{2,h}^n$ be the solutions of the systems (2.39) and (2.46), respectively. Then*

$$\|G_{1,h}(t_n) - G_{1,h}^n\|_{L^2(\Omega)} \le C\tau(t_n^{-1}\|G_{1,h}(0)\|_{L^2(\Omega)}$$
$$+ t_n^{\alpha_2-1}\|G_{2,h}(0)\|_{L^2(\Omega)}),$$
$$\|G_{2,h}(t_n) - G_{2,h}^n\|_{L^2(\Omega)} \le C\tau(t_n^{\alpha_1-1}\|G_{1,h}(0)\|_{L^2(\Omega)}$$
$$+ t_n^{-1}\|G_{2,h}(0)\|_{L^2(\Omega)}).$$

Proof: We first consider the error estimates between $G_{1,h}^n$ and $G_{1,h}$. According to (2.44), the solutions of (2.39) in Laplace space can be reconstructed as

$$\begin{aligned}\tilde{G}_{1,h} &= z^{-1}H_2(z^{-\alpha_2}, z^{-\alpha_1}, A_{2,h}, A_{1,h})G_{1,h}(0) \\ &\quad + aH_1(z^{-\alpha_2}, z^{-\alpha_1}, A_{2,h}, A_{1,h})z^{-1-\alpha_2}G_{2,h}(0),\end{aligned} \tag{2.49}$$

which leads to

$$\begin{aligned}G_{1,h}(t) &= \frac{1}{2\pi\mathbf{i}} \int_{\Gamma_{\theta,\kappa}} e^{zt}z^{-1}H_2(z^{-\alpha_2}, z^{-\alpha_1}, A_{2,h}, A_{1,h})G_{1,h}(0)dz \\ &\quad + \frac{1}{2\pi\mathbf{i}} \int_{\Gamma_{\theta,\kappa}} e^{zt}az^{-1}H_1(z^{-\alpha_2}, z^{-\alpha_1}, A_{2,h}, A_{1,h}) \\ &\quad \cdot z^{-\alpha_2}G_{2,h}(0)dz.\end{aligned} \tag{2.50}$$

Combining (2.47) and (2.50) leads to

$$
G_{1,h}(t_n) - G_{1,h}^n
$$

$$
= \frac{1}{2\pi \mathbf{i}} \int_{\Gamma_{\theta,\kappa} \backslash \Gamma_{\theta,\kappa}^\tau} e^{z t_n} z^{-1} H_2(z^{-\alpha_2}, z^{-\alpha_1}, A_{2,h}, A_{1,h}) G_{1,h}(0) dz
$$

$$
+ \frac{1}{2\pi \mathbf{i}} \int_{\Gamma_{\theta,\kappa} \backslash \Gamma_{\theta,\kappa}^\tau} e^{z t_n} a z^{-1} H_1(z^{-\alpha_2}, z^{-\alpha_1}, A_{2,h}, A_{1,h}) z^{-\alpha_2} G_{2,h}(0) dz
$$

$$
+ \frac{1}{2\pi \mathbf{i}} \int_{\Gamma_{\theta,\kappa}^\tau} e^{z t_n} \left(z^{-1} H_2(z^{-\alpha_2}, z^{-\alpha_1}, A_{2,h}, A_{1,h}) \right.
$$

$$
\left. - e^{-z\tau} \left(\frac{1 - e^{-z\tau}}{\tau} \right)^{-1} \right.
$$

$$
\left. \cdot H_2(\psi^{(1-\alpha_2)}(e^{-z\tau}), \psi^{(1-\alpha_1)}(e^{-z\tau}), A_{2,h}, A_{1,h}) \right) G_{1,h}(0) dz
$$

$$
+ \frac{a}{2\pi \mathbf{i}} \int_{\Gamma_{\theta,\kappa}^\tau} e^{z t_n} \left(z^{-1} H_1(z^{-\alpha_2}, z^{-\alpha_1}, A_{2,h}, A_{1,h}) z^{-\alpha_2} \right.
$$

$$
\left. - e^{-z\tau} \left(\frac{1 - e^{-z\tau}}{\tau} \right)^{-1} \right.
$$

$$
\left. \cdot H_1(\psi^{(1-\alpha_2)}(e^{-z\tau}), \psi^{(1-\alpha_1)}(e^{-z\tau}), A_{2,h}, A_{1,h})\psi^{(1-\alpha_2)}(e^{-z\tau}) \right)
$$

$$
\cdot G_{2,h}(0) dz
$$

$$
= I + II + III + IV.
$$

By using Remark 2.7, one can get

$$
\|I\|_{L^2(\Omega)}
$$

$$
\leq C \int_{\Gamma_{\theta,\kappa} \backslash \Gamma_{\theta,\kappa}^\tau} e^{-C|z|t_n} |z|^{-1}
$$

$$
\|H_2(z^{-\alpha_2}, z^{-\alpha_1}, A_{2,h}, A_{1,h})\|_{L^2(\Omega) \to L^2(\Omega)}
$$

$$
|dz| \|G_{1,h}(0)\|_{L^2(\Omega)}
$$

$$
\leq C t_n^{-1} \tau \|G_{1,h}(0)\|_{L^2(\Omega)}.
$$

Similarly, the II satisfies

$$
\|II\|_{L^2(\Omega)} \leq C t_n^{\alpha_2 - 1} \tau \|G_{2,h}(0)\|_{L^2(\Omega)}.
$$

Next we can separate III into three parts, i.e.,

III

$$= \frac{1}{2\pi\mathbf{i}} \int_{\Gamma_{\theta,\kappa}^{\tau}} e^{zt_n} e^{-z\tau} \left(e^{z\tau} z^{-1} H_2(z^{-\alpha_2}, z^{-\alpha_1}, A_{2,h}, A_{1,h}) \right.$$

$$\left. - \left(\frac{1-e^{-z\tau}}{\tau}\right)^{-1} H_2(z^{-\alpha_2}, z^{-\alpha_1}, A_{2,h}, A_{1,h}) \right) G_{1,h}(0) dz$$

$$+ \frac{1}{2\pi\mathbf{i}} \int_{\Gamma_{\theta,\kappa}^{\tau}} e^{zt_n} e^{-z\tau} \left(\left(\frac{1-e^{-z\tau}}{\tau}\right)^{-1} H_2(z^{-\alpha_2}, z^{-\alpha_1}, A_{2,h}, A_{1,h}) \right.$$

$$\left. - \left(\frac{1-e^{-z\tau}}{\tau}\right)^{-1} H_2(\psi^{(1-\alpha_2)}(e^{-z\tau}), \psi^{(1-\alpha_1)}(e^{-z\tau}), A_{2,h}, A_{1,h}) \right)$$

$$\cdot G_{1,h}(0) dz$$

$$+ \frac{1}{2\pi\mathbf{i}} \int_{\Gamma_{\theta,\kappa}^{\tau}} e^{zt_n} e^{-z\tau} (e^{z\tau} - 1) z^{-1} H_2(z^{-\alpha_2}, z^{-\alpha_1}, A_{2,h}, A_{1,h}) dz$$

$$= III_1 + III_2 + III_3.$$

As for III_1, Remark 2.7 leads to

$$\left\| z^{-1} H_2(z^{-\alpha_2}, z^{-\alpha_1}, A_{2,h}, A_{1,h}) - \right.$$

$$\left. \left(\frac{1-e^{-z\tau}}{\tau}\right)^{-1} H_2(z^{-\alpha_2}, z^{-\alpha_1}, A_{2,h}, A_{1,h}) \right\|_{L^2(\Omega) \to L^2(\Omega)} \le C\tau.$$

Thus

$$\|III_1\|_{L^2(\Omega)} \le C\tau \int_{\Gamma_{\theta,\kappa}^{\tau}} e^{-C|z|t_{n-1}} |dz| \|G_{1,h}(0)\|_{L^2(\Omega)}$$

$$\le C t_n^{-1} \tau \|G_{1,h}(0)\|_{L^2(\Omega)}.$$

Introduce $H^{(a,b)}(z_1, z_2, A_1, A_2)$ as the a-th order derivative about z_1 and b-th order derivative about z_2. Using the facts

$$\|H_2^{(1,0)}(z^{-\alpha_2}, z^{-\alpha_1}, A_{2,h}, A_{1,h})\|_{L^2(\Omega) \to L^2(\Omega)} \le C|z|^{\alpha_2},$$

$$\|H_2^{(0,1)}(z^{-\alpha_2}, z^{-\alpha_1}, A_{2,h}, A_{1,h})\|_{L^2(\Omega) \to L^2(\Omega)} \le C|z|^{\alpha_1},$$

the mean value theorem, Remark 2.8 and the Lemma 2.15, one has

$$\left\| \left(\frac{1 - e^{-z\tau}}{\tau} \right)^{-1} H_2(z^{-\alpha_2}, z^{-\alpha_1}, A_{2,h}, A_{1,h}) \right.$$

$$- \left(\frac{1 - e^{-z\tau}}{\tau} \right)^{-1} H_2(\psi^{(1-\alpha_2)}(e^{-z\tau}), \psi^{(1-\alpha_1)}(e^{-z\tau}), A_{2,h}, A_{1,h})$$

$$\left. \vphantom{\frac{1}{2}} \right\|_{L^2(\Omega) \to L^2(\Omega)} \leq C\tau.$$

Thus

$$\|III_2\|_{L^2(\Omega)} \leq C\tau \int_{\Gamma_{\theta,\kappa}^\tau} e^{-C|z|t_{n-1}} |dz| \|G_{1,h}(0)\|_{L^2(\Omega)}$$

$$\leq Ct_n^{-1} \tau \|G_{1,h}(0)\|_{L^2(\Omega)}.$$

Similarly, we have

$$\|III_3\|_{L^2(\Omega)} \leq C\tau \int_{\Gamma_{\theta,\kappa}^\tau} e^{-C|z|t_{n-1}} |dz| \|G_{1,h}(0)\|_{L^2(\Omega)}$$

$$\leq Ct_n^{-1} \tau \|G_{1,h}(0)\|_{L^2(\Omega)},$$

which leads to

$$\|III\|_{L^2(\Omega)} \leq Ct_n^{-1} \tau \|G_{1,h}(0)\|_{L^2(\Omega)}.$$

Similarly, one has

$$\|IV\|_{L^2(\Omega)} \leq Ct_n^{\alpha_2-1} \tau \|G_{2,h}(0)\|_{L^2(\Omega)}.$$

In summary,

$$\|G_{1,h}(t_n) - G_{1,h}^n\|_{L^2(\Omega)}$$

$$\leq C\tau \left(t_n^{-1} \|G_{1,h}(0)\|_{L^2(\Omega)} + t_n^{\alpha_2-1} \|G_{2,h}(0)\|_{L^2(\Omega)} \right).$$

Analogously, we get the error estimate of $\|G_{2,h}(t_n) - G_{2,h}^n\|_{L^2(\Omega)}$. $\square$

2.2.4 Numerical Experiments

In this section, the one-dimensional numerical experiments are provided to validate the effectiveness of the designed schemes. In Examples 2.5–2.8, due to the exact solutions G_1 and G_2 are unknown, the spatial error can be calculated by

$$E_{1,h} = \|G_{1,h}^n - G_{1,h/2}^n\|_{L^2(\Omega)}, \quad E_{2,h} = \|G_{2,h}^n - G_{2,h/2}^n\|_{L^2(\Omega)},$$

where $G_{1,h}^n$ and $G_{2,h}^n$ mean the numerical solutions of G_1 and G_2 at time t_n with mesh size h; similarly, the temporal error can be calculated by

$$E_{1,\tau} = \|G_{1,\tau} - G_{1,\tau/2}\|_{L^2(\Omega)}, \quad E_{2,\tau} = \|G_{2,\tau} - G_{2,\tau/2}\|_{L^2(\Omega)},$$

where $G_{1,\tau}$ and $G_{2,\tau}$ are the numerical solutions of G_1 and G_2 at the fixed time t with time step size τ.

In our numerical experiments, we consider the following two sets of conditions about initial values and domains

- $\quad G_{1,0}(x) = \chi_{(1/2,1)}, \quad G_{2,0}(x) = \chi_{(0,1/2)}, \quad \Omega = (0,1);$
  $$\tag{2.51}$$

- $\quad G_{1,0}(x) = (1-x)^{-\nu_1}, \quad G_{2,0}(x) = x^{-\nu_2}, \quad \Omega = (0,1).$
  $$\tag{2.52}$$

Here $\chi_{(a,b)}$ is the characteristic function on (a,b).

We first show the influence of the regularity of initial data on convergence rates by some examples.

Example 2.5 *We choose $s = s_1 = s_2 < 1/2$, $\alpha_1 = 0.4$, $\alpha_2 = 0.7$, $a = 2$, $\tau = 1/800$, and $T = 1$ to solve the system (2.31) with the condition (2.51). It is easy to get $G_{1,0}, G_{2,0} \in \hat{H}^{1/2-\epsilon}(\Omega)$, which satisfy the conditions of Theorem 2.9. We show the convergence rates in Table 2.6, which can be achieved as $\mathcal{O}(h^{s+1/2-\epsilon})$.*

Example 2.6 *We take $\alpha_1 = 0.4$, $\alpha_2 = 0.6$, $a = 2$, $\tau = 1/800$, and $T = 1$ to solve the system (2.31) with the condition (2.51). We present the L^2 errors and convergence rates for different values of s_1, s_2 in Table 2.7. When $s_1, s_2 > 1/2$, the convergence rates are consistent with the results of Theorem 2.9; but for $s_1, s_2 < 1/2$, the convergence rates of G_2 are higher than the predicted ones in Theorem 2.10 and the convergence rates of G_1 are the same as the predicted ones, which is due to the less effect of $aH(z, A_{1,h}, \alpha_1, \alpha_1 - \alpha_2)P_h\tilde{G}_2 - aH(z, A_{1,h}, \alpha_1, \alpha_1 - \alpha_2)\tilde{G}_{2,h}$ and $aH(z, A_{2,h}, \alpha_2, \alpha_2 - \alpha_1)P_h\tilde{G}_1 - aH(z, A_{2,h}, \alpha_2, \alpha_2 - \alpha_1)\tilde{G}_{1,h}$ in (2.41) on convergence rates.*

Example 2.7 *In this example, we choose $\alpha_1 = 0.7$, $\alpha_2 = 0.6$, $a = 0.1$, $\tau = 1/50$, and $T = 20$. We take the condition (2.52) with $\nu_1 = 0$, $\nu_2 = 0.4999$ as initial values, which implies $G_{1,0} \in$*

Table 2.6 L^2 errors and convergence rates with $s_1 = s_2 = s < 1/2$ and the condition (2.51)

s	$1/h$	50	100	200	400	800
0.1	$E_{1,h}$	1.293E-02	8.631E-03	5.752E-03	3.829E-03	2.546E-03
	Rate		0.5834	0.5854	0.5872	0.5888
	$E_{2,h}$	9.139E-03	6.038E-03	3.986E-03	2.630E-03	1.734E-03
	Rate		0.5981	0.5991	0.5999	0.6005
0.25	$E_{1,h}$	5.861E-03	3.486E-03	2.071E-03	1.229E-03	7.298E-04
	Rate		0.7496	0.7514	0.7522	0.7523
	$E_{2,h}$	3.795E-03	2.247E-03	1.331E-03	7.890E-04	4.680E-04
	Rate		0.7562	0.7553	0.7544	0.7535
0.4	$E_{1,h}$	2.334E-03	1.247E-03	6.681E-04	3.585E-04	1.925E-04
	Rate		0.9044	0.9006	0.8982	0.8971
	$E_{2,h}$	1.468E-03	7.863E-04	4.218E-04	2.264E-04	1.216E-04
	Rate		0.9010	0.8985	0.8974	0.8975

Table 2.7 L^2 errors and convergence rates with different s_1, s_2 and the condition (2.51)

(s_1, s_2)	$1/h$	50	100	200	400	800
(0.1,0.2)	$E_{1,h}$	1.173E-02	7.797E-03	5.181E-03	3.441E-03	2.284E-03
	Rate		0.5894	0.5899	0.5905	0.5913
	$E_{2,h}$	6.456E-03	3.988E-03	2.460E-03	1.516E-03	9.340E-04
	Rate		0.6949	0.6969	0.6982	0.6992
(0.3,0.4)	$E_{1,h}$	4.105E-03	2.349E-03	1.345E-03	7.707E-04	4.420E-04
	Rate		0.8051	0.8045	0.8034	0.8023
	$E_{2,h}$	1.853E-03	9.921E-04	5.320E-04	2.856E-04	1.534E-04
	Rate		0.9017	0.8989	0.8973	0.8968
(0.6,0.7)	$E_{1,h}$	5.780E-04	2.754E-04	1.326E-04	6.425E-05	3.109E-05
	Rate		1.0695	1.0547	1.0453	1.0472
	$E_{2,h}$	2.347E-04	1.092E-04	5.172E-05	2.479E-05	1.193E-05
	Rate		1.1032	1.0785	1.0613	1.0546
(0.8,0.9)	$E_{1,h}$	1.143E-04	4.905E-05	2.194E-05	1.013E-05	4.798E-06
	Rate		1.2207	1.1607	1.1149	1.0780
	$E_{2,h}$	3.297E-05	1.330E-05	5.799E-06	2.668E-06	1.269E-06
	Rate		1.3098	1.1974	1.1202	1.0714

$\hat{H}^{1/2-\epsilon}(\Omega)$, $G_{2,0} \in L^2(\Omega)$. *According to Table 2.8, the results for* $s_1 = 0.25$ *and* $s_2 = 0.8$ *agree with Theorem 2.10; when* s_1, $s_2 < 1/2$, *the convergence rates of* G_1 *are higher than the predicted ones in Theorem 2.10 and the convergence rates of* G_2 *are the same as the predicted ones.*

Table 2.8 L^2 errors and convergence rates with different s_1, s_2 and the condition (2.52) ($\nu_1 = 0$, $\nu_2 = 0.4999$)

(s_1, s_2)	$1/h$	50	100	200	400	800
(0.4,0.1)	$E_{1,h}$	3.630E-04	1.980E-04	1.080E-04	5.894E-05	3.217E-05
	Rate		0.8746	0.8742	0.8739	0.8737
	$E_{2,h}$	2.591E-02	2.269E-02	1.985E-02	1.736E-02	1.517E-02
	Rate		0.1916	0.1926	0.1936	0.1947
(0.4,0.2)	$E_{1,h}$	3.417E-04	1.849E-04	1.001E-04	5.411E-05	2.924E-05
	Rate		0.8858	0.8862	0.8869	0.8882
	$E_{2,h}$	1.122E-02	8.629E-03	6.622E-03	5.072E-03	3.878E-03
	Rate		0.3784	0.3818	0.3848	0.3874
(0.6,0.3)	$E_{1,h}$	8.932E-05	4.412E-05	2.196E-05	1.100E-05	5.464E-06
	Rate		1.0175	1.0065	0.9974	1.0095
	$E_{2,h}$	4.877E-03	3.321E 03	2.252E-03	1.521E-03	1.024E-03
	Rate		0.5544	0.5606	0.5660	0.5707

Table 2.9 L^2 errors and convergence rates with different α_1, α_2 and the condition (2.52)

(α_1, α_2)	$1/\tau$	100	200	400	800
(0.4,0.6)	$E_{1,\tau}$	2.2893E-04	1.1725E-04	5.9651E-05	3.0209E-05
	Rate		0.9653	0.9749	0.9816
	$E_{2,\tau}$	9.2000E-04	4.6012E-04	2.3015E-04	1.1513E-04
	Rate		0.9997	0.9994	0.9994
(0.3,0.7)	$E_{1,\tau}$	1.1613E-04	6.2013E-05	3.2538E-05	1.6874E-05
	Rate		0.9051	0.9304	0.9474
	$E_{2,\tau}$	1.1276E-03	5.6202E-04	2.8052E-04	1.4012E-04
	Rate		1.0046	1.0025	1.0014

Next, we verify the temporal convergence rates.

Example 2.8 *Here we take $s_1 = 0.25$, $s_2 = 0.75$, $a = 2$, $T = 1$ and $h = 1/400$ to solve the system (2.31) with the condition (2.52) and $\nu_1 = 0$, $\nu_2 = 0.49$. Table 2.9 shows the L^2 errors and convergence rates for different α_1, α_2, which can be used to validate the results of Theorem 2.11.*

CHAPTER 3

Numerical Methods for the Stochastic Governing Equations of PDF of Statistical Observables

As aforementioned, the deterministic dynamics with nonlocal operators is successfully used to explain phenomena and fit data in complex environments. But, it doesn't cover all the cases. In addition to the inherent laws of particle motion, sometimes the external stochastic disturbances can also affect the system. In this chapter, the stochastic models with Itô noise have been studied including stochastic partial differential equations (SPDEs) and stochastic wave equations (SWEs).

Over the last few decades, there is much progress in both strong and weak approximations of the stochastic model equations. In [52,192], the finite element approximation of some linear stochastic PDEs is studied in space. The numerical schemes of the semilinear or nonlinear SPDEs are discussed [23,93,119,121,193,195]. The work [103] investigates a discrete approximation of the linear SPDE with a positive-type memory. A full discretization of the SWE driven by additive space-time white noise is presented with a spectral Galerkin approximation in space and a temporal approximation by exponential time integrators involving linear functionals of the white noise [181]. In [11, 39], the stochastic trigonometric method for solving the SWE with multiplicative space-time white noise is studied in time. The work [115] investigates a discrete approximation for the stochastic space-time fractional wave equation forced by an additive space-time white noise. To obtain the higher order approximation of stochastic space fractional wave equation, the authors of [120] design a fully discrete scheme which consists of a temporal approximation by modifying the stochastic trigonometric method and postprocessing spectral Galerkin method in space.

The models discussed in this chapter involve in the fractional Laplacian in the spectral sense, white noise, and fractional Gaussian noise. In Section 3.1, we first review some basic properties of stochastic processes which are used to estimates regularity and error. We prove the existence and uniqueness of the mild solution to SPDEs forced by a tempered fractional Gaussian noise in Section 3.2. Moreover, the Hölder continuity of the mild solution is discussed. Then, the spectral Galerkin method is used for space approximation; after that the system is transformed into an equivalent form having better regularity than the original one in time. And we use the semi-implicit Euler scheme to discretize the time derivative. In Section 3.3, we establish the regularity of the mild solution of the stochastic fractional wave equations involving space-fractional operators and time-fractional derivatives. We design numerical methods for stochastic fractional wave equations. And the higher order approximation for stochastic space fractional wave equation is presented.

3.1 ASSUMPTIONS AND GAUSSIAN PROCESSES

We gather preliminary results on the Dirichlet eigenpairs and fBm, which are commonly used in the chapter.

Let $U = L^2(D; \mathbb{R})$ be a real separable Hilbert space with L^2 inner product $\langle \cdot, \cdot \rangle$ and the corresponding induced norm $\|\cdot\|$. We define the unbounded linear operator A^ν by $A^\nu u = (-\Delta)^\nu u$ on the domain

$$\text{dom}\,(A^\nu) = \{A^\nu u \in U : u(x) = 0,\ x \in \partial D\}.$$

Then

$$A^{\frac{\nu}{2}} \phi_i(x) = \lambda_i^{\frac{\nu}{2}} \phi_i(x)$$

and

$$A^{\frac{\nu}{2}} u = \sum_{i=1}^{\infty} \lambda_i^{\frac{\nu}{2}} \langle u, \phi_i(x) \rangle \, \phi_i(x),$$

where $\phi_i(x)$, $i = 1, 2, \ldots$, denote the normalized eigenfunctions of the fractional Laplacian operator $(-\Delta)^{\frac{\nu}{2}}$, and $\lambda_i^{\frac{\nu}{2}}$, $i = 1, 2, \ldots$, are the corresponding eigenvalues. Moreover, we define the Hilbert space $\dot{U}^\nu = \mathrm{dom}\left(A^{\frac{\nu}{2}}\right)$ equipped with the inner product

$$\langle u, v \rangle_\nu = \sum_{i=1}^{\infty} \lambda_i^{\frac{\nu}{2}} \langle u, \phi_i(x) \rangle \times \lambda_i^{\frac{\nu}{2}} \langle v, \phi_i(x) \rangle$$

and norm

$$\|u\|_\nu = \left(\sum_{i=1}^{\infty} \lambda_i^{\nu} \langle u, \phi_i(x) \rangle^2 \right)^{\frac{1}{2}}.$$

In particular, $\dot{U}^0 = U$.

Lemma 3.1 ([109, 114, 170]) *Let Ω denote a bounded domain in $\mathbb{R}^d$, $d \in \{1, 2, 3\}$. Let λ_i be the i-th eigenvalue of the homogeneous Dirichlet boundary problem for the Laplacian operator $-\Delta$ in Ω. Then*

$$C_0 i^{\frac{2}{d}} \leq \lambda_i \leq C_1 i^{\frac{2}{d}},$$

where $i \in \mathbb{N}$, and the constants C_0 and C_1 are independent of i.

Assumption 1 *The function $f : U \to U$ satisfies*

$$\|f(u) - f(v)\| \lesssim \|u - v\| \text{ for any } u, v \in U,$$

and

$$\|A^{\frac{\nu}{2}} f(u)\| \lesssim 1 + \|A^{\frac{\nu}{2}} u\| \text{ for } u \in \dot{U}^\nu \text{ with } \nu \geq 0.$$

For later use, we collect concepts of Gaussian process; for more details, one can refer to [9, 18, 22, 48, 53, 74, 138].

Fractional Brownian motion (fBm) is a Gaussian process, which has two important properties (self similarity and stationary increments), and has long-range correlation. The integral representation of fractional Brownian motion is as follows:

$$\beta_H(t) = \frac{1}{\Gamma(\frac{1}{2} + H)} \int_{-\infty}^{+\infty} \left[(t - y)_+^{H - \frac{1}{2}} - (-y)_+^{H - \frac{1}{2}} \right] d\beta(y),$$

where $\beta(t)$ is a standard one dimensional Brownian motion (Bm), Hurst index $H \in (0,1)$, $\Gamma(y)$ is the gamma function, and

$$(y)_+ = \begin{cases} y, & y > 0, \\ 0, & y \le 0. \end{cases}$$

Using the following methods [42, 81, 82, 143, 163], we can define fBm on a finite interval, that is

$$\beta_H(t) = \int_0^t K_H(t,s)\beta(s), \quad t > 0.$$

As $\frac{1}{2} < H < 1$,

$$K_H(t,s) = c_H \left(H - \frac{1}{2} \right) s^{\frac{1}{2}-H} \int_s^t (u-s)^{H-\frac{3}{2}} u^{H-\frac{1}{2}} \mathrm{d}u,$$

where $t > s$ and

$$c_H = \left(\frac{2H \times \Gamma(\frac{3}{2} - H)}{\Gamma(\frac{1}{2} + H)\Gamma(2 - 2H)} \right)^{\frac{1}{2}}.$$

For $0 < H < \frac{1}{2}$,

$$K_H(t,s) = c_H(t-s)^{H-\frac{1}{2}} + c_H \left(\frac{1}{2} - H \right)$$
$$\int_s^t (u-s)^{\frac{3}{2}-H} \left(1 - \left(\frac{s}{u} \right)^{\frac{1}{2}-H} \right) \mathrm{d}u.$$

Definition 3.1 *Let $\beta_H(t)$ be the two-sided one-dimensional fBm with Hurst index $H \in (0,1)$ and $t \in \mathbb{R}$. The stochastic process $\beta_H(t)$ is characterized by the properties:*
(i) $\beta_H(0) = 0$;
(ii) $\mathrm{E}[\beta_H(t)] = 0$, $t \in \mathbb{R}$;
(iii) $\mathrm{E}[\beta_H(t)\beta_H(s)] = \frac{1}{2}(|t|^{2H} + |s|^{2H} - |t-s|^{2H})$, $t, s \in \mathbb{R}$,

where E denotes the expectation. As $H = \frac{1}{2}$, $\beta_H(t)$ is a standard Bm, being a process with independent increment.

The theory of stochastic integral for fBm was developed in [22, 41, 54, 74, 138, 178]. If $\frac{1}{2} < H < 1$ and $(K_t^* f)(s) \in L^2([0,T], \mathbb{R})$, the definition of stochastic integral for fBm is

$$\int_0^t f(s)\mathrm{d}\beta_H(s) = \int_0^t (K_t^* f)(s)\mathrm{d}\beta(s), \quad s, t \in [0,T],$$

where

$$(K_t^* f)(s) = \int_s^t f(r) \frac{\partial K_H(r,s)}{\partial r} \mathrm{d}r$$

$$= c_H \left(H - \frac{1}{2} \right) \int_s^t f(r)(r-s)^{H-\frac{3}{2}} \left(\frac{s}{r} \right)^{\frac{1}{2}-H} \mathrm{d}r.$$

Lemma 3.2 *For $f, g \in L^2(\mathbb{R}; \mathbb{R}) \cap L^1(\mathbb{R}; \mathbb{R})$, as $\frac{1}{2} < H < 1$, we have*

$$\mathrm{E} \left[\int_R f(s) \mathrm{d}\beta_H(s) \right] = 0$$

and

$$\mathrm{E} \left[\int_R f(s) \mathrm{d}\beta_H(s) \int_R g(t) \mathrm{d}\beta_H(t) \right]$$

$$= H(2H-1) \int_R \int_R [f(s)g(t)] |s-t|^{2H-2} \mathrm{d}s \mathrm{d}t.$$

Lemma 3.2 implies

$$\mathrm{E} \left[\mathrm{d}\beta_H(t) \mathrm{d}\beta_H(s) \right] = H(2H-1)|t-s|^{2H-2} \mathrm{d}t \mathrm{d}s.$$

Next, we introduce the infinite dimensional tempered fractional Brownian motion (tfBm). The tfBm, describing anomalous diffusion with exponentially tempered long range correlations, self similarity and stationary increments, is still a Gaussian process. The definition of tfBm is given as

$$\beta_{H,\mu}(t) = \int_{-\infty}^{+\infty} \left[e^{-\mu(t-y)_+}(t-y)_+^{H-\frac{1}{2}} - e^{-\mu(-y)_+}(-y)_+^{H-\frac{1}{2}} \right] \mathrm{d}\beta(y),$$

where $H > 0, H \neq \frac{1}{2}, \mu > 0$. For a more detailed introduction of tfBm, see [37, 131, 132].

The expectation and covariance function of tfBm, respectively, are

$$\mathrm{E} \left[\beta_{H,\mu}(t) \right] = 0$$

and

$$\mathrm{E} \left[\beta_{H,\mu}(t)\beta_{H,\mu}(s) \right] = \frac{1}{2} \left[C_t^2 |t|^{2H} + C_s^2 |s|^{2H} - C_{t-s}^2 |t-s|^{2H} \right],$$

where

$$C_t^2 = \int_{-\infty}^{+\infty} \left[e^{-\mu|t|(1-y)_+}(1-y)_+^{H-\frac{1}{2}} - e^{-\mu|t|(-y)_+}(-y)_+^{H-\frac{1}{2}} \right]^2 \mathrm{d}y$$

$$= \frac{2\Gamma(2H)}{(2\mu|t|)^{2H}} - \frac{2\Gamma\left(H+\frac{1}{2}\right)K_H(\mu|t|)}{\sqrt{\pi}(2\mu|t|)^H}$$

for $t \neq 0$ and $C_0^2 = 0$, $\Gamma(y)$ is the gamma function, and $K_H(y)$ is the modified Bessel function of the second kind

$$K_H(y) = \frac{1}{2} \int_0^{+\infty} t^{H-1} \exp\left[-\frac{1}{2}y\left(t + \frac{1}{t}\right)\right] dt.$$

The theory of the stochastic integration for tfBm was developed in [132, 157, 162]. If $H > \frac{1}{2}$, $\mu > 0$ and $f(y) \in L^2(\mathbb{R}; \mathbb{R})$, the stochastic integral is

$$\int_{\mathbb{R}} f(y)d\beta_{H,\mu}(y) = \Gamma(H + \frac{1}{2}) \int_{\mathbb{R}} \left[\mathbb{I}^{H-\frac{1}{2},\mu} f(y) - \mu \mathbb{I}^{H+\frac{1}{2},\mu} f(y)\right] d\beta(y),$$

(3.1)

where

$$\mathbb{I}^{H,\mu} f(y) = \frac{1}{\Gamma(H)} \int_y^{+\infty} f(x)(x-y)^{H-1} e^{-\mu(x-y)} dx.$$

Let $W^{\nu,2}$ be Sobolev space. For $0 < H < \frac{1}{2}$, $\mu > 0$ and $f(y) \in W^{\frac{1}{2}-H,2}(\mathbb{R}; \mathbb{R})$, the stochastic integral is defined as

$$\int_{\mathbb{R}} f(y)d\beta_{H,\mu}(y) = \Gamma(H + \frac{1}{2}) \int_{\mathbb{R}} \left[\mathbb{D}^{\frac{1}{2}-H,\mu} f(y) - \mu \mathbb{I}^{H+\frac{1}{2},\mu} f(y)\right] d\beta(y),$$

(3.2)

where

$$\mathbb{D}^{H,\mu} f(y) = \mu^H f(y) + \frac{H}{\Gamma(1-H)} \int_y^{+\infty} \frac{f(y) - f(x)}{(x-y)^{H+1}} e^{-\mu(x-y)} dx.$$

Assumption 2 *Let driven stochastic process $G(x,t)$ be a cylindrical Gaussian process with respect to the normal filtration $\{\mathcal{F}_t\}_{t\in[0,T]}$. The infinite dimensional space-time stochastic process can be represented by the formal series*

$$G(x,t) = \sum_{i=1}^{\infty} \sigma_i g^i(t)\phi_i(x),$$

where $|\sigma_i| \lesssim \lambda_i^{-\rho}$ ($\rho \geq 0$, λ_i is given in Lemma 3.1), $g^i(t)$, $i = 1, 2, \ldots$, are mutually independent real-valued Gaussian processes, and $\{\phi_i(x)\}_{i\in\mathbb{N}}$ is an orthonormal basis of U.

We define $L^p(D, \dot{U}^\nu)$ to be the separable Hilbert space of p-times integrable random variables with norm

$$\|u\|_{L^p(D,\dot{U}^\nu)} = (\mathbb{E}[\|u\|_\nu^p])^{\frac{1}{p}}, \quad \nu \geq 0.$$

We shall frequently use the Mittag-Leffler function $E_{\alpha,\beta}(z)$ defined as follows:

$$E_{\alpha,\beta}(z) = \sum_{i=0}^{\infty} \frac{z^i}{\Gamma(i\alpha + \beta)}, \quad z \in \mathbb{C}.$$

The Mittag-Leffler function $E_{\alpha,\beta}(z)$ is a two-parameter family of entire functions in z of order α^{-1} and type 1 [96]. It generalizes the exponential function in the sense that $E_{\alpha,\beta}(z) = e^z$. For later use, we collect some results in the next lemma; see [96, 150].

Lemma 3.3 *Let $0 < \alpha < 2$ and $\beta \in \mathbb{R}$ be arbitrary. We suppose that μ is an arbitrary real number such that $\frac{\pi\alpha}{2} < \mu < \min\{\pi, \pi\alpha\}$. Then there exists a constant $C = C(\alpha, \beta, \mu) > 0$ such that*

$$|E_{\alpha,\beta}(z)| \leq \frac{C}{1 + |z|}, \quad \mu \leq |\arg(z)| \leq \pi.$$

Moreover, for $\lambda > 0$, $\alpha > 0$, and positive integer $m \in \mathbb{N}$, we have

$$\frac{\mathrm{d}^m}{\mathrm{d}t^m} E_{\alpha,1}(-\lambda^\beta t^\alpha) = -\lambda^\beta t^{\alpha-m} E_{\alpha,1}(-\lambda^\beta t^\alpha), \quad t > 0,$$

and

$$\frac{\mathrm{d}}{\mathrm{d}t}\left(t E_{\alpha,2}(-\lambda^\beta t^\alpha)\right) = E_{\alpha,1}(-\lambda^\beta t^\alpha), \quad t \geq 0.$$

Definition 3.2 *The fractional integral of order $\alpha > 0$ with the lower limit 0 for a function u is defined as*

$$I_t^\alpha u(t) = \frac{1}{\Gamma(\alpha)} \int_0^t (t-s)^{\alpha-1} u(s)\mathrm{d}s, \quad t > 0,$$

where $\Gamma(\alpha)$ is the gamma function.

Definition 3.3 *The Caputo fractional derivative of order $\alpha > 0$ with the lower limit 0 for a function u is defined as*

$$\partial_t^\alpha u(t) = \frac{1}{\Gamma(n-\alpha)} \int_0^t (t-s)^{n-\alpha-1} \frac{\partial^n u(s)}{\partial s^n} \mathrm{d}s, \quad t > 0,$$
$$0 \leq n-1 < \alpha < n,$$

where the function $u(t)$ has absolutely continuous derivatives up to order $n - 1$.

3.2 NUMERICAL SCHEMES FOR STOCHASTIC FRACTIONAL DIFFUSION EQUATION

3.2.1 Numerical Approximation of Stochastic Fractional Diffusion Equation with a Tempered Fractional Gaussian Noise

The "localization" is one of the typical properties of fluctuation, well modeled by the tempered fractional Brownian motion (tfBm) [37, 132]. Let us consider the following model with the fluctuation source term (tempered fractional Gaussian noise) and the deterministic source term $f(u(x,t))$ depending on the concentration of the particles

$$\frac{\partial u(x,t)}{\partial t} = -(-\Delta)^\alpha u(x,t) + f\left(u(x,t)\right) + \dot{B}_{H,\mu}(x,t) \qquad (3.3)$$

with the initial and boundary conditions given by

$$u(x,0) = u_0(x), \quad x \in D,$$
$$u(x,t) = 0, \quad (x,t) \in \partial D \times [0,T],$$
$$\text{and } \alpha \in (0,1).$$

Here $\dot{B}_{H,\mu}(x,t) = \frac{\partial B_{H,\mu}(x,t)}{\partial t}$ is the tempered fractional Gaussian noise; the definition of infinite dimensional tfBm $B_{H,\mu}(x,t)$ is given in Assumption 2 and H is Hurst parameter.

3.2.1.1 Regularity of the solution

We give the formal mild solution of Eq. (3.3) and prove the existence and uniqueness of the mild solution (the presentation is for the two dimensional case). Moreover, the Hölder continuity of the mild solution is discussed. These results will be used for numerical analysis.

For the sake of brevity, we rewrite Eq. (3.3) as

$$\begin{cases} du(t) + A^\alpha u(t)dt = f\left(u(t)\right)dt + dB_{H,\mu}(t), & \text{in } D \times (0,T], \\ u(0) = u_0, & \text{in } D, \\ u(t) = 0, & \text{on } \partial D, \end{cases}$$
$$(3.4)$$

where $u(t) = u(x,t)$ and $B_{H,\mu}(t) = B_{H,\mu}(x,t)$. There is a formal mild solution $u(t)$ for Eq. (3.4), that is

$$u(t) = S(t)u_0 + \int_0^t S(t-s)f\left(u(s)\right)ds + \int_0^t S(t-s)dB_{H,\mu}(s), \quad (3.5)$$

where $S(t) = e^{-tA^\alpha}$. Equation (3.5) shows a fact that besides the initial value u_0, the regularity of the mild solution for Eq. (3.3) depends on the stochastic integration $\int_0^t S(t - s) \mathrm{d}B_{H,\mu}(s)$. Therefore, we need to obtain the following estimates in the first place.

Proposition 3.1 Let $0 \leq \tilde{t} < t$, $0 < \epsilon < 2$, $\gamma = 2\rho - 1 + (2 - \epsilon)\alpha \cdot \min\{H, 1\}$, and $\rho > \frac{1}{2} - \frac{(2-\epsilon)\alpha}{2} \cdot \min\{H, 1\}$. Then

$$\mathrm{E}\left[\left\|A^{\frac{\gamma}{2}} \int_{\tilde{t}}^t S(t - s) \mathrm{d}B_{H,\mu}(s)\right\|^2\right] \lesssim \frac{(t - \tilde{t})^{2H - \frac{(4-\epsilon)\cdot\min\{H,1\}}{2}}}{\alpha\,(\epsilon \cdot \min\{H, 1\})^2}.$$

Proof: As $H > \frac{1}{2}$, by using Eq. (3.1) and triangle inequality, we have

$$\mathrm{E}\left[\left\|A^{\frac{\gamma}{2}} \int_{\tilde{t}}^t S(t - s) \mathrm{d}B_{H,\mu}(s)\right\|^2\right]$$

$$\lesssim \mathrm{E}\left[\left\|\sum_i \int_{\tilde{t}}^t \frac{\lambda_i^{\frac{\gamma}{2}} \sigma_i \phi_i(x)}{\Gamma(H - \frac{1}{2})}\right.\right.$$

$$\left.\left. \times \int_s^t e^{-\lambda_i^\alpha(t-u)}(u - s)^{H - \frac{3}{2}} e^{-\mu(u-s)} \mathrm{d}u \mathrm{d}\beta^i(s)\right\|^2\right]$$

$$+ \mathrm{E}\left[\left\|\sum_i \int_{\tilde{t}}^t \frac{\lambda_i^{\frac{\gamma}{2}} \sigma_i \mu \phi_i(x)}{\Gamma(H + \frac{1}{2})}\right.\right.$$

$$\left.\left. \times \int_s^t e^{-\lambda_i^\alpha(t-u)}(u - s)^{H - \frac{1}{2}} e^{-\mu(u-s)} \mathrm{d}u \mathrm{d}\beta^i(s)\right\|^2\right]$$

$$= J_1 + J_2.$$

As $\theta_1 < 1$, the integration $\int_s^t (t - u)^{-\theta_1}(u - s)^{H - \frac{3}{2}} \mathrm{d}u$ is finite. Let $\theta = \frac{(4-\epsilon)\cdot\min\{H,1\}}{4}$. Then combining Lemma 3.1, Itô's isometry, $e^{-x} \lesssim x^{-\theta_1}(x, \theta_1 \geq 0)$, and the mutual independence of $\beta^i(s)$, we have

$$
\begin{aligned}
J_1 &= \sum_i \int_{\tilde{t}}^{t} \left[\frac{\lambda_i^{\frac{\gamma}{2}} \sigma_i}{\Gamma(H - \frac{1}{2})} \int_s^t e^{-\lambda_i^{\alpha}(t-u)} (u-s)^{H-\frac{3}{2}} e^{-\mu(u-s)} du \right]^2 ds \\
&\lesssim \sum_i \int_{\tilde{t}}^{t} \left[\lambda_i^{\frac{\gamma}{2}} \lambda_i^{-\rho} \int_s^t e^{-\lambda_i^{\alpha}(t-u)} (u-s)^{H-\frac{3}{2}} du \right]^2 ds \\
&\lesssim \sum_i \int_{\tilde{t}}^{t} \lambda_i^{-1-\frac{\epsilon\alpha\cdot\min\{H,1\}}{2}} \left[\int_s^t (t-u)^{-\theta} (u-s)^{H-\frac{3}{2}} du \right]^2 ds \\
&\lesssim \sum_i \int_{\tilde{t}}^{t} i^{-1-\frac{\epsilon\alpha\cdot\min\{H,1\}}{2}} (t-s)^{2H-2\theta-1} ds \\
&\lesssim \frac{\left(t-\tilde{t}\right)^{2H-\frac{(4-\epsilon)\cdot\min\{H,1\}}{2}}}{\alpha\left(\epsilon\cdot\min\{H,1\}\right)^2}.
\end{aligned}
\tag{3.6}
$$

For J_2, similarly, we have

$$
J_2 \lesssim \frac{\left(t-\tilde{t}\right)^{2H+2-\frac{(4-\epsilon)\cdot\min\{H,1\}}{2}}}{\alpha\epsilon\cdot\min\{H,1\}}.
\tag{3.7}
$$

As $0 < H < \frac{1}{2}$, by using Eq. (3.1), we have

$$
\begin{aligned}
&\mathrm{E}\left[\left\|A^{\frac{\gamma}{2}} \int_{\tilde{t}}^{t} S(t-s) \mathrm{d}B_{H,\mu}(s)\right\|^2\right] \\
&\lesssim \mathrm{E}\left[\left\|\sum_i \int_{\tilde{t}}^{t} \lambda_i^{\frac{\gamma}{2}} \sigma_i \phi_i(x) e^{-\lambda_i^{\alpha}(t-s)} \mathrm{d}\beta^i(s)\right\|^2\right] \\
&+ \mathrm{E}\left[\left\|\sum_i \int_{\tilde{t}}^{t} \lambda_i^{\frac{\gamma}{2}} \sigma_i \phi_i(x)\right.\right. \\
&\qquad \left.\left.\times \int_s^t \frac{e^{-\lambda_i^{\alpha}(t-s)} - e^{-\lambda_i^{\alpha}(t-u)}}{(u-s)^{\frac{3}{2}-H}} e^{-\mu(u-s)} \mathrm{d}u \mathrm{d}\beta^i(s)\right\|^2\right] \\
&+ \mathrm{E}\left[\left\|\sum_i \int_{\tilde{t}}^{t} \lambda_i^{\frac{\gamma}{2}} \sigma_i \mu \phi_i(x)\right.\right. \\
&\qquad \left.\left.\times \int_s^t e^{-\lambda_i^{\alpha}(t-u)} (u-s)^{H-\frac{1}{2}} e^{-\mu(u-s)} \mathrm{d}u \mathrm{d}\beta^i(s)\right\|^2\right] \\
&= \tilde{J}_1 + \tilde{J}_2 + \tilde{J}_3.
\end{aligned}
$$

Similar to the derivation of Eq. (3.6), we have

$$\tilde{J}_1 \lesssim \sum_i \lambda_i^{\gamma-2\rho-\alpha}\left(1 - e^{-2\lambda_i^\alpha\left(t-\tilde{t}\right)}\right)$$

$$\lesssim \frac{1}{\alpha(1-2H)}. \tag{3.8}$$

For the term $\tilde{J}_2$, by using the fact that $e^{-x} - e^{-y} \lesssim |x-y|^{\theta_1}$ with $x, y \geq 0$ and $0 \leq \theta_1 \leq 1$, we get

$$\tilde{J}_2 \lesssim \sum_i \int_{\tilde{t}}^t \lambda_i^{\gamma-2\rho}\left(\int_s^t (\lambda_i^\alpha(u-s))^\delta \frac{(\lambda_i^\alpha(t-u))^{-\eta}}{(u-s)^{\frac{3}{2}-H}}\,du\right)^2 ds$$

$$\lesssim \sum_i \lambda_i^{-1-\frac{\epsilon\alpha H}{2}} \frac{\left(t-\tilde{t}\right)^{\frac{\epsilon H}{2}}}{\epsilon H}$$

$$\lesssim \frac{\left(t-\tilde{t}\right)^{\frac{\epsilon H}{2}}}{\alpha\left(\epsilon H\right)^2}. \tag{3.9}$$

In the second inequality, we choose $\delta > \frac{1}{2} - H$ and $\eta < 1$ such that the integration $\int_s^t (u-s)^{\delta+H-\frac{3}{2}}(t-u)^{-\eta}du$ is bounded and $\eta \quad \delta = \frac{(4-\epsilon)H}{4}$. For the term $\tilde{J}_3$, we have

$$\tilde{J}_3 \lesssim \frac{\left(t-\tilde{t}\right)^{2+\frac{\epsilon H}{2}}}{\alpha\epsilon H}. \tag{3.10}$$

Then combining Eqs. (3.6)–(3.10) leads to

$$\mathrm{E}\left[\left\|A^{\frac{\gamma}{2}}\int_{\tilde{t}}^t S(t-s)\mathrm{d}B_{H,\mu}(s)\right\|^2\right] \lesssim \frac{\left(t-\tilde{t}\right)^{2H-\frac{(4-\epsilon)\cdot\min\{H,1\}}{2}}}{\alpha\left(\epsilon\cdot\min\{H,1\}\right)^2},$$

which completes the proof. □

Remark 3.1 *As $\rho \geq 0.5$, Proposition 3.1 holds for any α and H. If $\alpha H > 0.5$ and $\alpha > 0.5$, one can choose $\rho = 0$.*

The following theorem shows the regularity results of the mild solution of Eq. (3.4) by using Proposition 3.1 and Dirichlet eigenpairs $\{(\lambda_i, \phi_i(x))\}_{i\in\mathbb{N}}$.

Theorem 3.1 *Suppose that Assumptions 1 and 2 are satisfied,* $\|u(0)\|_{L^2(D,\dot{U}^\gamma)} < \infty$, $0 < \epsilon < 2$, $\gamma = 2\rho - 1 + (2 - \epsilon)\alpha \cdot \min\{H, 1\}$, *and* $\rho > \frac{1}{2} - \frac{2-\epsilon}{2}\alpha \cdot \min\{H, 1\}$. *Then Eq. (3.3) possesses a unique mild solution*

$$\|u(t)\|_{L^2(D,\dot{U}^\gamma)} \lesssim \frac{\alpha^{-\frac{1}{2}}}{\epsilon \cdot \min\{H, 1\}} + \|u_0\|_{L^2(D,\dot{U}^\gamma)}.$$

Moreover, we have
(i) For $\rho > \frac{1}{2}$,

$$\|u(t) - u(s)\|_{L^2(D,U)}$$
$$\lesssim (t - s)^{\min\{H,1\}} \left(\frac{1}{\min\{\alpha H, 2\rho - 1\}} + \|u_0\|_{L^2(D,\dot{U}^\gamma)} \right);$$

(ii) For $\frac{1}{2} - \frac{2-\epsilon}{2}\alpha \cdot \min\{H, 1\} < \rho \leq \frac{1}{2}$,

$$\|u(t) - u(s)\|_{L^2(D,U)} \lesssim (t - s)^{\frac{\gamma}{2\alpha}} \left(\frac{1}{\epsilon \cdot \min\{H, 1\}} + \|u_0\|_{L^2(D,\dot{U}^\gamma)} \right).$$

Proof: By using Proposition 3.1 and triangle inequality, the regularity property of the mild solution $u(t)$ can be established as

$$\mathrm{E}\left[\left\|A^{\frac{\gamma}{2}}u(t)\right\|^2\right] \lesssim \mathrm{E}\left[\left\|A^{\frac{\gamma}{2}}u(0)\right\|^2\right]$$
$$+ \int_0^t \mathrm{E}\left[\left\|A^{\frac{\gamma}{2}}u(s)\right\|^2\right] ds + \frac{1}{\alpha \left(\epsilon \cdot \min\{H, 1\}\right)^2}.$$
$$(3.11)$$

Then the Grönwall inequality leads to

$$\|u(t)\|_{L^2(D,\dot{U}^\gamma)} \lesssim \frac{\alpha^{-\frac{1}{2}}}{\epsilon \cdot \min\{H, 1\}} + \|u_0\|_{L^2(D,\dot{U}^\gamma)}. \qquad (3.12)$$

Meanwhile, we prove the Hölder continuity of the mild solution $u(t)$. Equation (3.5) implies that

$$
\mathrm{E}\left[\|u(t) - u(s)\|^2\right]
$$

$$
\lesssim \mathrm{E}\left[\|(S(t) - S(s))\,u(0)\|^2\right] + \mathrm{E}\left[\left\|\int_s^t S(t - r)\mathrm{d}B_{H,\mu}(r)\right\|^2\right]
$$

$$
+ \mathrm{E}\left[\left\|\int_0^s \left(S(t - r) - S(s - r)\right)\mathrm{d}B_{H,\mu}(r)\right\|^2\right]
$$

$$
+ \mathrm{E}\left[\left\|\int_s^t S(t - r)f\left(u(r)\right)\mathrm{d}r\right\|^2\right]
$$

$$
+ \mathrm{E}\left[\left\|\int_0^s \left(S(t - r) - S(s - r)\right)f\left(u(r)\right)\mathrm{d}r\right\|^2\right]
$$

$$
- \quad I_1 + I_2 + I_3 + I_4 + I_5.
$$

Let $\theta = \min\left\{\frac{\gamma}{2\alpha}, 1\right\}$. Then we have

$$
I_1 \; = \; \mathrm{E}\left[\left\|\sum_i \left(e^{-\lambda_i^\alpha t} - e^{-\lambda_i^\alpha s}\right)\langle u(0), \phi_i(x)\rangle\,\phi_i(x)\right\|^2\right]
$$

$$
\lesssim \; \mathrm{E}\left[\left\|\sum_i (t - s)^\theta \lambda_i^{\frac{\gamma}{2}}\langle u(0), \phi_i(x)\rangle\,\phi_i(x)\right\|^2\right]
$$

$$
\lesssim \; (t - s)^{\min\left\{\frac{\gamma}{\alpha}, 2\right\}}\mathrm{E}\left[\left\|A^{\frac{\gamma}{2}}u(0)\right\|^2\right]. \tag{3.13}
$$

In the first inequality, we have used the fact $\gamma \geq 2\alpha$, when $\theta = 1$.

For I_4, we need to estimate the upper bound of $\mathrm{E}\left[\|u(t)\|^2\right]$. Let $\theta_1 = \min\{H, 1\} - \frac{\min\left\{H, \frac{\gamma}{2\alpha}, 1\right\}}{2}$. Then Proposition 3.1 implies

$$
\mathrm{E}\left[\left\|\int_{\tilde{t}}^t S(t-r)\mathrm{d}B_{H,\mu}(r)\right\|^2\right]
$$

$$
\lesssim \sum_i \int_{\tilde{t}}^t \left[\frac{\sigma_i}{\Gamma(H - \frac{1}{2})} \int_r^t e^{-\lambda_i^\alpha(t-u)}(u-r)^{H-\frac{3}{2}}e^{-\mu(u-r)}\mathrm{d}u\right]^2 \mathrm{d}r
$$

$$
\lesssim \sum_i \int_{\tilde{t}}^t \left[\lambda_i^{-\rho} \int_r^t e^{-\lambda_i^\alpha(t-u)}(u-r)^{H-\frac{3}{2}}\mathrm{d}u\right]^2 \mathrm{d}r
$$

$$
\lesssim \sum_i \int_{\tilde{t}}^t \lambda_i^{-2\rho - 2\alpha\theta_1} \left[\int_r^t (t-u)^{-\theta_1}(u-r)^{H-\frac{3}{2}}\mathrm{d}u\right]^2 \mathrm{d}r
$$

$$
\lesssim \sum_i \int_{\tilde{t}}^t \lambda_i^{-\gamma-1+\alpha\cdot\min\left\{H, \frac{\gamma}{2\alpha}, 1\right\}}(t-r)^{2H-2\theta_1-1}\mathrm{d}r
$$

$$
\lesssim \frac{1}{\gamma \cdot \min\{2\alpha H, \gamma\}}.
$$

Finally, similar to Eq. (3.11), we have

$$
\mathrm{E}\left[\|u(t)\|^2\right] \lesssim \frac{1}{\gamma \cdot \min\{2\alpha H, \gamma\}} + \mathrm{E}\left[\|u(0)\|^2\right].
$$

Combining the Hölder inequality and Assumption 1 leads to

$$
\begin{aligned}
I_4 &= \mathrm{E}\left[\left\|\sum_i \int_s^t e^{-\lambda_i^\alpha(t-r)}\langle f(u(r)), \phi_i(x)\rangle \phi_i(x)\mathrm{d}r\right\|^2\right] \\
&\lesssim \mathrm{E}\left[\left\|\int_s^t f(u(r))\mathrm{d}r\right\|^2\right] \\
&\lesssim (t-s)\mathrm{E}\left[\int_s^t (\|u(r)\| + 1)^2 \mathrm{d}r\right] \\
&\lesssim (t-s)^2 \left(\frac{1}{\gamma \cdot \min\{2\alpha H, \gamma\}} + \mathrm{E}\left[\|u(0)\|^2\right]\right).
\end{aligned}
$$

For I_5, similar to I_4, there is

$$
\mathrm{E}\left[\left\|A^{\frac{\gamma}{4}}u(t)\right\|^2\right] \lesssim \frac{1}{\gamma \cdot \min\{2\alpha H, \gamma\}} + \mathrm{E}\left[\left\|A^{\frac{\gamma}{4}}u(0)\right\|^2\right]. \qquad (3.14)
$$

Let $\theta_2 = \min\left\{\frac{\gamma}{2\alpha}, 1\right\}$. Then

$$I_5 = \mathrm{E}\left[\left\|\sum_i \int_0^s \left(e^{-\lambda_i^\alpha(t-s)} - 1\right) e^{-\lambda_i^\alpha(s-r)} \langle f(u(r)), \phi_i(x)\rangle \phi_i(x)\mathrm{d}r\right\|^2\right]$$

$$\lesssim (t-s)^{2\theta_2}\mathrm{E}\left[\left\|\sum_i \int_0^s \lambda_i^{\frac{\alpha\theta_2}{2}}(s-r)^{-\frac{\theta_2}{2}} \langle f(u(r)), \phi_i(x)\rangle \phi_i(x)\mathrm{d}r\right\|^2\right]$$

$$\lesssim (t-s)^{2\theta_2}\mathrm{E}\left[\left\|\int_0^s (s-r)^{-\frac{\theta_2}{2}} A^{\frac{\gamma}{4}} f(u(r))\mathrm{d}r\right\|^2\right]$$

$$\lesssim (t-s)^{2\theta_2}\mathrm{E}\left[\left(\int_0^s (s-r)^{-\frac{\theta_2}{2}} \left(\left\|A^{\frac{\gamma}{4}}u(r)\right\| + 1\right)\mathrm{d}r\right)^2\right]$$

$$\lesssim (t-s)^{2\theta_2}\mathrm{E}\left[\int_0^s (s-r)^{-\frac{\theta_2}{2}} \left(\left\|A^{\frac{\gamma}{4}}u(r)\right\| + 1\right)^2 \mathrm{d}r \int_0^s (s-r)^{-\frac{\theta_2}{2}}\mathrm{d}r\right]$$

$$\lesssim (t-s)^{\min\left\{\frac{\gamma}{\alpha}, 2\right\}} \left(\frac{1}{\gamma \cdot \min\{2\alpha H, \gamma\}} + \mathrm{E}\left[\left\|A^{\frac{\gamma}{4}}u(0)\right\|^2\right]\right).$$

As $\rho > \frac{1}{2}$, Proposition 3.1 leads to

$$I_2 = \mathrm{E}\left[\left\|\int_s^t S(t-r)\mathrm{d}B_{H,\mu}(r)\right\|^2\right] \lesssim \frac{(t-s)^{2H}}{(2\rho-1)H}. \quad (3.15)$$

Let $\theta_3 = \min\{H, 1\}$, $\theta_4 = \theta_3 - \min\left\{\frac{2\rho-1}{4\alpha}, \frac{\theta_3}{2}\right\}$, and $H > \frac{1}{2}$. Using Eq. (3.1) and Proposition 3.1 leads to

$$\begin{aligned}
I_3 &\lesssim (t-s)^{2\theta_3} \sum_i \int_0^s \lambda_i^{2\alpha\theta_3}\sigma_i^2 \left(\int_r^s e^{-\lambda_i^\alpha(s-u)}(u-r)^{H-\frac{3}{2}}\mathrm{d}u\right)^2 \mathrm{d}r \\
&+ (t-s)^{2\theta_3} \sum_i \int_0^s \lambda_i^{2\alpha\theta_3}\sigma_i^2 \left(\int_r^s e^{-\lambda_i^\alpha(s-u)}(u-r)^{H-\frac{1}{2}}\mathrm{d}u\right)^2 \mathrm{d}r \\
&\lesssim (t-s)^{\min\{2H,2\}} \sum_i i^{2\alpha(\theta_3-\theta_4)-2\rho}\frac{1}{H-\theta_4} \\
&\lesssim \frac{(t-s)^{\min\{2H,2\}}}{(2\rho-1)\min\{(2\rho-1), \alpha H\}}.
\end{aligned}$$

As $0 < H < \frac{1}{2}$, by using the same procedure, we have $I_2 + I_3 \lesssim \frac{(t-s)^{\min\{2H,2\}}}{(2\rho-1)H}$. Then

$$\begin{aligned}
&\|u(t) - u(s)\|_{L^2(D,U)} \\
&\lesssim (t-s)^{\min\{H,1\}} \left(\frac{1}{\min\{\alpha H, 2\rho-1\}} + \|u_0\|_{L^2(D,\dot{U}^\gamma)}\right).
\end{aligned}$$

For $\frac{1}{2} - \frac{2-\epsilon}{2}\alpha \cdot \min\{H,1\} < \rho \leq \frac{1}{2}$, as $H > \frac{1}{2}$, Proposition 3.1 leads to

$$
\begin{aligned}
I_2 &= \mathrm{E}\left[\left\|\int_s^t S(t-r)\mathrm{d}B_{H,\mu}(r)\right\|^2\right] \\
&\lesssim \sum_i \lambda_i^{-(1+\epsilon\alpha\cdot\min\{H,1\})} \\
&\quad \times \int_s^t \left(\int_r^t (t-u)^{-\frac{1-2\rho+\epsilon\alpha\cdot\min\{H,1\}}{2\alpha}}(u-r)^{H-\frac{3}{2}}\mathrm{d}u\right)^2 \mathrm{d}r \\
&\lesssim \sum_i i^{-1-\epsilon\alpha\cdot\min\{H,1\}} \int_s^t (t-r)^{-\frac{1-2\rho+\epsilon\alpha\cdot\min\{H,1\}}{\alpha}+2H-1}\mathrm{d}r \\
&\lesssim \frac{1}{\epsilon\alpha H\gamma}(t-s)^{\frac{2\alpha H-1+2\rho-\epsilon\alpha\cdot\min\{H,1\}}{\alpha}}.
\end{aligned}
\tag{3.16}
$$

In the second inequality, we have used the fact that $2\alpha > 1 - 2\rho + \epsilon\alpha \cdot \min\{H,1\}$. In the fourth inequality, we have used the fact that $\gamma = 2\rho - 1 + (2-\epsilon)\alpha \cdot \min\{H,1\} > 0$. For I_3, the condition $\rho \leq \frac{1}{2}$ leads to $\gamma < 2\alpha$. Thus we have

$$
\begin{aligned}
I_3 &\lesssim (t-s)^{\frac{\gamma}{\alpha}} \sum_i \int_0^s \lambda_i^{\gamma-2\rho-\frac{(4-\epsilon)\alpha\cdot\min\{H,1\}}{2}} \\
&\quad \times \left(\int_r^s (s-u)^{-\frac{(4-\epsilon)\cdot\min\{H,1\}}{4}}(u-r)^{H-\frac{3}{2}}\mathrm{d}u\right)^2 \mathrm{d}r \\
&\lesssim (t-s)^{\frac{\gamma}{\alpha}} \sum_i \lambda_i^{-1-\frac{\epsilon\alpha\cdot\min\{H,1\}}{2}} \frac{1}{\epsilon\cdot\min\{H,1\}} \\
&\lesssim \frac{(t-s)^{\frac{\gamma}{\alpha}}}{(\epsilon\cdot\min\{H,1\})^2}.
\end{aligned}
\tag{3.17}
$$

Similarly, for $0 < H < \frac{1}{2}$, we have $I_2 + I_3 \lesssim \frac{(t-s)^{\frac{\gamma}{\alpha}}}{(\epsilon\cdot\min\{H,1\})^2}$. Due to $2\alpha H - 1 + 2\rho - \epsilon\alpha \cdot \min\{H,1\} \geq \gamma$, then

$$
\|u(t) - u(s)\|_{L^2(D,U)} \lesssim (t-s)^{\frac{\gamma}{2\alpha}}\left(\frac{1}{\epsilon\cdot\min\{H,1\}} + \|u_0\|_{L^2(D,\dot{U}^\gamma)}\right),
$$

which completes the proof. □

3.2.1.2 Galerkin approximation for spatial discretization

We provide the Galerkin spatial semi-discretization of Eq. (3.3). The error estimates are also presented. To implement the Galerkin spatial approximation of Eq. (3.3), we choose a finite dimensional subspace of U. Let U^N be a N dimensional subspace of U, and the sequence $\{\phi_1(x), \ldots, \phi_i(x), \ldots, \phi_N(x)\}_{N \in \mathbb{N}}$ is an orthonormal basis of U^N. Then we introduce the projection operator $P_N : U \to U^N$: for $\xi \in U$,

$$P_N\xi = \sum_{i=1}^{N} \langle \xi, \phi_i(x) \rangle \phi_i(x)$$

and

$$\langle P_N\xi, \chi \rangle = \langle \xi, \chi \rangle \quad \forall \chi \in U^N. \tag{3.18}$$

Additionally, define $A_N^\alpha : U^N \to U^N$ by

$$A_N^\alpha \chi = A^\alpha P_N \chi = P_N A^\alpha \chi = \sum_{i=1}^{N} \lambda_i^\alpha \langle \chi, \phi_i(x) \rangle \phi_i(x) \quad \forall \chi \in U^N. \tag{3.19}$$

The Galerkin formulation of Eq. (3.3) is: Find $u^N(t) \in U^N$ such that

$$\begin{cases} \langle du^N(t), \chi \rangle + \langle A^\alpha u^N(t)dt, \chi \rangle - \langle f(u^N(t))\, dt, \chi \rangle + \langle dB_{H,\mu}(t), \chi \rangle, \\ \langle u^N(0), \chi \rangle = \langle u(0), \chi \rangle, \end{cases} \tag{3.20}$$

where $\chi \in U^N$. Then according to Eqs. (3.18), (3.19), and (3.20), the Galerkin approximation of Eq. (3.3) is obtained in the weak sense

$$\begin{cases} du^N(t) + A_N^\alpha u^N(t)dt = f_N(u^N(t))\, dt + P_N dB_{H,\mu}(t), & t \in (0, T], \\ u^N(0) = P_N u(0), \end{cases} \tag{3.21}$$

where $f_N = P_N f$. Similar to the Eq. (3.1), the unique mild solution of Eq. (3.21) is given by

$$\begin{aligned} u^N(t) &= S_N(t)u_0 + \int_0^t S_N(t-s)f_N(u^N(s))\, ds \\ &\quad + \int_0^t S_N(t-s)dB_{H,\mu}(s), \end{aligned} \tag{3.22}$$

where $S_N(t) = e^{-tA_N^\alpha}$. Theorem 3.1 implies the following result.

Corollary 3.1 *Suppose that Assumptions 1 and 2 are satisfied,* $\|u(0)\|_{L^2(D, \dot{U}^\gamma)} < \infty$, $0 < \epsilon < 2$, $\gamma = 2\rho - 1 + (2 - \epsilon)\alpha \cdot \min\{H, 1\}$,

$\rho > \frac{1}{2} - \frac{(2-\epsilon)}{2}\alpha \cdot \min\{H, 1\}$, *and $u^N(t)$ is the unique mild solution of Eq. (3.21). Then*

$$\left\|u^N(t)\right\|_{L^2(D,\dot{U}^\gamma)} \lesssim \frac{\alpha^{-\frac{1}{2}}}{\epsilon \cdot \min\{H, 1\}} + \|u_0\|_{L^2(D,\dot{U}^\gamma)}$$

and we obtain the Hölder regularity of the mild solution $u^N(t)$:
(i) *For $\rho > \frac{1}{2}$,*

$$\left\|u^N(t) - u^N(s)\right\|_{L^2(D,U)}$$
$$\lesssim (t-s)^{\min\{H,1\}} \left(\frac{1}{\min\{\alpha H, 2\rho - 1\}} + \|u_0\|_{L^2(D,\dot{U}^\gamma)}\right);$$

(ii) *For $\frac{1}{2} - \frac{(2-\epsilon)}{2}\alpha \cdot \min\{H, 1\} < \rho \le \frac{1}{2}$,*

$$\left\|u^N(t) - u^N(s)\right\|_{L^2(D,\dot{U}^\gamma)} \lesssim (t-s)^{\frac{\gamma}{2\alpha}}$$
$$\times \left(\frac{1}{\epsilon \cdot \min\{H, 1\}} + \|u_0\|_{L^2(D,\dot{U}^\gamma)}\right).$$

To analyze the error of the Galerkin spatial semi-discretization for Eq. (3.21), the following lemmas are needed.

Lemma 3.4 *If $\mathrm{E}\left[\|A^{\frac{\nu}{2}}\xi\|^2\right] < \infty$, $\xi \in U$, then*

$$\mathrm{E}\left[\|(P_N - I)\xi\|^2\right] \lesssim \lambda_{N+1}^{-\nu} \mathrm{E}\left[\|A^{\frac{\nu}{2}}\xi\|^2\right].$$

Proof:

$$\mathrm{E}\left[\|(P_N - I)\xi\|^2\right] = \mathrm{E}\left[\left\|\sum_{i=N+1}^{\infty} \langle\xi, \phi_i(x)\rangle \phi_i(x)\right\|^2\right]$$
$$\lesssim \lambda_{N+1}^{-\nu}\mathrm{E}\left[\left\|\sum_{i=N+1}^{\infty} \lambda_i^{\frac{\nu}{2}} \langle\xi, \phi_i(x)\rangle \phi_i(x)\right\|^2\right]$$
$$\lesssim \lambda_{N+1}^{-\nu}\mathrm{E}\left[\|A^{\frac{\nu}{2}}\xi\|^2\right].$$

□

From Lemma 3.4, one can infer the bound of the spatial error for scheme (3.4) in the $L^2(D, U)$.

Theorem 3.2 *Let $u(t)$ and $u^N(t)$ be, respectively, the mild solutions of Eq. (3.4) and Eq. (3.21) with the assumptions given in Theorem 3.1 and Corollary 3.1. Then we have*

$$\left\| u(t) - u^N(t) \right\|_{L^2(D,U)} \lesssim \lambda_{N+1}^{-\frac{\gamma}{2}} \left(\frac{\alpha^{-\frac{1}{2}}}{\epsilon \cdot \min\{H, 1\}} + \|u_0\|_{L^2(D,\dot{U}^\gamma)} \right).$$

Proof: Using the triangle inequality, Lemma 3.2, and Theorem 3.1, we obtain

$$
\begin{aligned}
& \mathrm{E}\left[\|u(t) - u^N(t)\|^2 \right] \\
\lesssim\ & \mathrm{E}\left[\|u(t) - P_N u(t)\|^2 \right] + \mathrm{E}\left[\|P_N u(t) - u^N(t)\|^2 \right] \\
\lesssim\ & \lambda_{N+1}^{-\gamma} \left(\frac{1}{\alpha\,(\epsilon \cdot \min\{H, 1\})^2} + \mathrm{E}\left[\left\| A^{\frac{\gamma}{2}} u(0) \right\|^2 \right] \right) \\
& + \mathrm{E}\left[\left\| P_N u(t) - u^N(t) \right\|^2 \right].
\end{aligned}
$$

Then it is needed to estimate the bound of $\mathrm{E}\left[\|P_N u(t) - u^N(t)\|^2 \right]$. Let $e_t^N = P_N u(t) - u^N(t)$. First performing P_N on Eq. (3.4) and then doing subtraction with respect to (3.22) lead to

$$e_t^N = \int_0^t S_N(t - s) \left[f(u(s)) - f_N(u^N(s)) \right] \mathrm{d}s.$$

Then

$$\frac{\mathrm{d}}{\mathrm{d}t} e_t^N = -A_N^\alpha e_t^N + f(u(t)) - f_N(u^N(t)),$$

which implies

$$
\begin{aligned}
\frac{\mathrm{d}}{\mathrm{d}t} \left\| e_t^N \right\|^2 &= 2 \left\langle e_t^N, -A_N^\alpha e_t^N + f(u(t)) - f\left(u^N(t)\right) \right\rangle \\
&\lesssim -\left\| A_N^{\frac{\alpha}{2}} e_t^N \right\|^2 + \left\| e_t^N \right\| \left\| u(t) - u^N(t) \right\| \\
&\lesssim \left\| e_t^N \right\| \left\| u(t) - P_N u(t) + P_N u(t) - u^N(t) \right\| \\
&\lesssim \left\| e_t^N \right\|^2 + \left\| e_t^N \right\| \| u(t) - P_N u(t) \| \\
&\lesssim \left\| e_t^N \right\|^2 + \| u(t) - P_N u(t) \|^2 ; \qquad (3.23)
\end{aligned}
$$

the Hölder inequality is used for the first inequality, and the fact that $ab \leq \frac{a^2}{2C} + \frac{Cb^2}{2}$ is used for the fourth inequality. Integrating Eq. (3.23) from 0 to t, from Theorem 3.1 and Lemma 3.4, we have

$$
\begin{aligned}
\mathrm{E}\left[\left\|e_t^N\right\|^2\right] &\lesssim \int_0^t \mathrm{E}\left[\left\|e_s^N\right\|^2 + \|u(s) - P_N u(s)\|^2\right] \mathrm{d}s \\
&\lesssim \int_0^t \mathrm{E}\left[\left\|e_s^N\right\|^2\right] \mathrm{d}s \\
&\quad + \lambda_{N+1}^{-\gamma}\left(\frac{1}{\alpha\left(\epsilon \cdot \min\{H, 1\}\right)^2} + \mathrm{E}\left[\left\|A^{\frac{\gamma}{2}} u(0)\right\|^2\right]\right).
\end{aligned}
$$

By using the Grönwall inequality, it has

$$
\mathrm{E}\left[\left\|e_t^N\right\|^2\right] \lesssim \lambda_{N+1}^{-\gamma}\left(\frac{1}{\alpha\left(\epsilon \cdot \min\{H, 1\}\right)^2} + \mathrm{E}\left[\left\|A^{\frac{\gamma}{2}} u(0)\right\|^2\right]\right).
$$

Then

$$
\left\|u(t) - u^N(t)\right\|_{L^2(D,U)} \lesssim \lambda_{N+1}^{-\frac{\gamma}{2}}\left(\frac{\alpha^{-\frac{1}{2}}}{\epsilon \cdot \min\{H, 1\}} + \|u_0\|_{L^2(D,\dot{U}^\gamma)}\right).
$$

$\square$

Note that if N is big enough, taking $\epsilon = \frac{1}{\log(\lambda_{N+1})}$ leads to

$$
\begin{aligned}
&\left\|u(t) - u^N(t)\right\|_{L^2(D,U)} \\
&\lesssim \lambda_{N+1}^{-\rho+\frac{1}{2}-\alpha\cdot\min\{H,1\}}\left(\frac{\alpha^{-\frac{1}{2}} \cdot \log(\lambda_{N+1})}{\min\{H, 1\}} + \|u_0\|_{L^2(D,\dot{U}^\gamma)}\right).
\end{aligned}
$$

3.2.1.3 *Fully discrete scheme*

We are concerned with the time discretization of Eq. (3.21). Meanwhile, the error estimates for the fully discrete scheme are derived.

Using the semi-implicit Euler scheme, one can get the fully discrete scheme of Eq. (3.3) as

$$
u_{m+1}^N - u_m^N + \tau A_N^\alpha u_{m+1}^N = \tau f_N\left(u_m^N\right) + P_N\left(B_{H,\mu}(t_{m+1}) - B_{H,\mu}(t_m)\right),
$$
$$
(3.24)
$$

but Proposition 3.1 implies that as $\alpha \geq \gamma$, $A^{\frac{\alpha}{2}} u(t)$ is not Hölder continuous, i.e.,

$$
\lim_{s \to t}\left\|A^{\frac{\alpha}{2}}\left(u(t) - u(s)\right)\right\|_{L^2(D,U)} \neq 0.
$$

Then the approximation scheme (3.24) is invalid. Therefore, we introduce the following technique to circumvent this defect.

Let $z^N(t) = u^N(t) - \int_0^t S_N(t-s)P_N dB_{H,\mu}(s)$. If $u^N(t)$ is the unique mild solution of Eq. (3.21), then $z^N(t)$ is the unique mild solution of the following PDE

$$\frac{\mathrm{d}}{\mathrm{d}t} z^N(t) + A_N^\alpha z^N(t) = f_N\left(u^N(t)\right) \quad \text{with } t \in (0,T]$$
$$\text{and } z^N(0) = u^N(0).$$

(3.25)

The unique mild solution of Eq. (3.25) is given by

$$z^N(t) = S_N(t)z^N(0) + \int_0^t S_N(t-s)f\left(u^N(s)\right) \mathrm{d}s.$$

If $\lim_{s \to t} \left\| A_N^{\frac{\alpha}{2}}(z(t) - z(s)) \right\|_{L^2(D,U)} = 0$, then one can use Euler scheme to obtain the time discretization of Eq. (3.25). The following theorem shows that $A^{\frac{\alpha}{2}}z(t)$ is Hölder continuous.

Theorem 3.3 *Let Assumptions 1 and 2 be fulfilled and $z^N(t)$ be the mild solution of Eq. (3.25). Let $\|u(0)\|_{L^2(D,\dot{U}^{\gamma+\alpha})} < \infty$ and the conditions of Corollary 3.1 are also satisfied. Then we have*

$$\left\| A_N^{\frac{\alpha}{2}}\left(z^N(t) - z^N(s)\right) \right\|_{L^2(D,U)}$$
$$\lesssim (t-s)^{\min\{\frac{\gamma}{2\alpha},1\}} \left(\frac{1}{\min\{\gamma\alpha H, \gamma^2\}} + \|u(0)\|_{L^2(D,\dot{U}^{\gamma+\alpha})} \right).$$

Proof: By using the inequality $(a+b)^2 \lesssim a^2 + b^2$, we get

$$\mathrm{E}\left[\left\| A_N^{\frac{\alpha}{2}}\left(z^N(t) - z^N(s)\right) \right\|^2\right]$$
$$\lesssim \mathrm{E}\left[\left\| A_N^{\frac{\alpha}{2}}(S_N(t) - S_N(s))z^N(0) \right\|^2\right]$$
$$+ \mathrm{E}\left[\left\| \int_s^t A_N^{\frac{\alpha}{2}}S_N(t-r)f\left(u^N(r)\right) \mathrm{d}r \right\|^2\right]$$
$$+ \mathrm{E}\left[\left\| \int_0^s A_N^{\frac{\alpha}{2}}(S_N(t-r) - S_N(s-r))f\left(u^N(r)\right) \mathrm{d}r \right\|^2\right]$$
$$= \tilde{I}_1 + \tilde{I}_2 + \tilde{I}_3.$$

(3.26)

As $2\alpha > \gamma$, the inequality $e^{-x} - e^{-y} \lesssim |x - y|^{\theta_1}$ $(0 \le \theta_1 \le 1, x \ge 0, y \ge 0)$ implies

$$
\begin{aligned}
\tilde{I}_1 &= \mathrm{E}\left[\left\|\sum_{i=1}^{N} \lambda_i^{\frac{\alpha}{2}} \left(e^{-\lambda_i^\alpha t} - e^{-\lambda_i^\alpha s}\right) \left\langle z^N(0), \phi_i(x) \right\rangle \phi_i(x)\right\|^2\right] \\
&\lesssim \mathrm{E}\left[\left\|\sum_{i=1}^{N} \lambda_i^{\frac{\alpha}{2}} \left((\lambda_i^\alpha (t - s))^{\frac{\gamma}{2\alpha}}\right) \left\langle z^N(0), \phi_i(x) \right\rangle \phi_i(x)\right\|^2\right] \\
&= \mathrm{E}\left[\left\|(t - s)^{\frac{\gamma}{2\alpha}} A_N^{\frac{\gamma+\alpha}{2}} z^N(0)\right\|^2\right] \\
&\lesssim (t - s)^{\frac{\gamma}{\alpha}} \mathrm{E}\left[\left\|A_N^{\frac{\gamma+\alpha}{2}} u^N(0)\right\|^2\right].
\end{aligned}
\tag{3.27}
$$

Combining the inequality $e^{-x} \lesssim x^{-\theta_1}$ $(\theta_1 \ge 0, x \ge 0)$ and Eq. (3.14) leads to

$$
\begin{aligned}
\tilde{I}_2 &\lesssim \mathrm{E}\left[\left\|\int_s^t \sum_{i=1}^{N} \lambda_i^{\frac{\alpha}{2}} (\lambda_i^\alpha (t - r))^{-\frac{\alpha - \frac{\gamma}{2}}{2\alpha}} \left\langle f\left(u^N(r)\right), \phi_i(x) \right\rangle \phi_i(x) \mathrm{d}r\right\|^2\right] \\
&= \mathrm{E}\left[\left\|\int_s^t (t - r)^{-\frac{\alpha - \frac{\gamma}{2}}{2\alpha}} A^{\frac{\gamma}{4}} f_N\left(u^N(r)\right) \mathrm{d}r\right\|^2\right] \\
&\lesssim \int_s^t \mathrm{E}\left[\left\|A^{\frac{\gamma}{4}} f_N\left(u^N(r)\right)\right\|^2\right] \mathrm{d}r \int_s^t (t - r)^{-\frac{\alpha - \frac{\gamma}{2}}{\alpha}} \mathrm{d}r \\
&\lesssim (t - s)^{1 + \frac{\gamma}{2\alpha}} \left(\frac{1}{\min\{\gamma\alpha H, \gamma^2\}} + \mathrm{E}\left[\left\|A^{\frac{\gamma}{4}} u(0)\right\|^2\right]\right).
\end{aligned}
\tag{3.28}
$$

Similar to the derivation of Eqs. (3.27) and (3.28), we have

$$
\begin{aligned}
\tilde{I}_3 &\lesssim \mathrm{E}\left[\left\|\int_0^s A_N^{\frac{\alpha}{2}} (A_N^\alpha (s - r))^{-\frac{1}{2} - \frac{\gamma}{4\alpha}} (A_N^\alpha (t - s))^{\frac{\gamma}{2\alpha}} f\left(u^N(r)\right) \mathrm{d}r\right\|^2\right] \\
&\lesssim (t - s)^{\frac{\gamma}{\alpha}} \left(\frac{1}{\min\{\gamma\alpha H, \gamma^2\}} + \mathrm{E}\left[\left\|A^{\frac{\gamma}{4}} u(0)\right\|^2\right]\right).
\end{aligned}
\tag{3.29}
$$

Combining Eqs. (3.27), (3.28), and (3.29) leads to

$$
\begin{aligned}
&\mathrm{E}\left[\left\|A_N^{\frac{\alpha}{2}} \left(z^N(t) - z^N(s)\right)\right\|^2\right] \\
&\lesssim (t - s)^{\frac{\gamma}{\alpha}} \left(\frac{1}{\min\{\gamma\alpha H, \gamma^2\}} + \mathrm{E}\left[\left\|A^{\frac{\gamma}{4}} u(0)\right\|^2\right]\right).
\end{aligned}
\tag{3.30}
$$

As $2\alpha \le \gamma$, by using the same procedure, we have

$$\mathrm{E}\left[\left\|A_N^{\frac{\alpha}{2}}\left(z^N(t) - z^N(s)\right)\right\|^2\right]$$

$$\lesssim (t-s)^2 \mathrm{E}\left[\left\|A_N^{\frac{3\alpha}{2}} u^N(0)\right\|^2\right] + \mathrm{E}\left[\left\|\int_s^t A_N^{\frac{\alpha}{2}} S_N(t-r) f\left(u^N(r)\right) dr\right\|^2\right]$$

$$+ \mathrm{E}\left[\left\|\int_0^s A_N^{\frac{\alpha}{2}}\left(S_N(s-r)\right) - \left(S_N(t-r)\right) f\left(u^N(r)\right) dr\right\|^2\right]$$

$$\lesssim (t-s)^2 \mathrm{E}\left[\left\|A_N^{\frac{3\alpha}{2}} u^N(0)\right\|^2\right] + \mathrm{E}\left[\left\|\int_s^t A_N^{\frac{\gamma}{4}} f\left(u^N(r)\right) dr\right\|^2\right]$$

$$+ (t-s)^2 \mathrm{E}\left[\left\|\int_0^s (s-r)^{-\frac{5}{6}} A_N^{\frac{2\alpha}{3}} f\left(u^N(r)\right) dr\right\|^2\right]$$

$$\lesssim (t-s)^2 \left(\frac{1}{\min\{\gamma\alpha H, \gamma^2\}} + \mathrm{E}\left[\left\|A_N^{\frac{\gamma+\alpha}{2}} u^N(0)\right\|^2\right]\right). \qquad (3.31)$$

Combining Eqs. (3.30) and (3.31) results in

$$\mathrm{E}\left[\left\|A_N^{\frac{\alpha}{2}}\left(z^N(t) - z^N(s)\right)\right\|^2\right]$$

$$\lesssim (t-s)^{\min\{\frac{\gamma}{\alpha}, 2\}} \left(\frac{1}{\min\{\gamma\alpha H, \gamma^2\}} + \mathrm{E}\left[\left\|A_N^{\frac{\gamma+\alpha}{2}} u^N(0)\right\|^2\right]\right).$$

$\square$

For time discretization of Eq. (3.25), we apply the classical semi-implicit Euler scheme. Let $\tau = \frac{T}{M}$ and $t_m = m\tau$ with $m = 0, 1, \ldots, M$. Then one can obtain an approximation $z_m^{N,M}$ of $z^N(t_m)$ by the recurrence

$$z_{m+1}^{N,M} - z_m^{N,M} + \tau A_N^\alpha z_{m+1}^{N,M} = \tau f_N\left(u_m^{N,M}\right), \qquad (3.32)$$

where $u_m^{N,M} = z_m^{N,M} + \int_0^{t_m} S_N(t_m - s) P_N dB_{H,\mu}(s)$. The approximation of $\int_0^{t_m} S_N(t_m - s) P_N dB_{H,\mu}(s)$ is given as

$$\int_0^{t_m} S_N(t_m - s) P_N dB_{H,\mu}(s)$$

$$\approx \sum_{k=0}^{t_m/\tilde{\tau}-1} S_N(t_m - k\tilde{\tau})\left(B_{H,\mu}((k+1)\tilde{\tau}) - B_{H,\mu}(k\tilde{\tau})\right).$$

Then

$$u_m^{N,M} = z_m^{N,M} + \sum_{k=0}^{t_m/\widetilde{\tau}-1} S_N(t_m - k\widetilde{\tau})\left(B_{H,\mu}((k+1)\widetilde{\tau}) - B_{H,\mu}(k\widetilde{\tau})\right).$$

(3.33)

For the sake of completeness, we need to derive the error estimates for the approximation of $\int_0^{t_m} S_N(t_m - s)P_N \mathrm{d}B_{H,\mu}(s)$.

Proposition 3.2 *Under the conditions of Proposition 3.1 and $\widetilde{\tau} < 1$, we have*

$$\left\|\sum_{k=0}^{t_m/\widetilde{\tau}-1} \int_{k\widetilde{\tau}}^{(k+1)\widetilde{\tau}} (S_N(t_m - s) - S_N(t_m - k\widetilde{\tau}))\,\mathrm{d}B_{H,\mu}(s)\right\|_{L^2(D,U)}$$

$$\lesssim \begin{cases} (2\rho H - H)^{-\frac{1}{2}}\frac{1}{(2\rho-1)^2}\widetilde{\tau}^H, & \rho > \frac{1}{2}, \\ (\epsilon\alpha H\gamma)^{-\frac{1}{2}}\frac{1}{\epsilon}\widetilde{\tau}^{H-\frac{1-2\rho+\epsilon\alpha\cdot\min\{H,1\}}{2\alpha}}, & \\ \frac{1}{2} - \frac{(2-\epsilon)}{2}\alpha\cdot\min\{H,1\} < \rho \le \frac{1}{2}. \end{cases}$$

Proof: As $0 < H < \frac{1}{2}$, Eq. (3.2) and the triangle inequality imply that

$$\left\|\sum_{k=0}^{t_m/\widetilde{\tau}-1} \int_{k\widetilde{\tau}}^{(k+1)\widetilde{\tau}} (S_N(t_m - s) - S_N(t_m - k\widetilde{\tau}))\,\mathrm{d}B_{H,\mu}(s)\right\|_{L^2(D,U)}$$

$$\lesssim \sum_{k=0}^{t_m/\widetilde{\tau}-1} \left(\mathrm{E}\left[\left\|\sum_{i=1}^{N} \int_{k\widetilde{\tau}}^{(k+1)\widetilde{\tau}} \left(e^{-(t_m-s)\lambda_i^\alpha}\right.\right.\right.\right.$$

$$\left.\left.\left.\left. - e^{-(t_m-k\widetilde{\tau})\lambda_i^\alpha}\right)\sigma_i\phi_i(x)\mathrm{d}\beta^i(s)\right\|^2\right]\right)^{\frac{1}{2}}$$

$$+ \sum_{k=0}^{t_m/\widetilde{\tau}-1} \left(\mathrm{E}\left[\left\|\sum_{i=1}^{N} \int_{k\widetilde{\tau}}^{(k+1)\widetilde{\tau}} \int_s^{(k+1)\widetilde{\tau}} \left(e^{-(t_m-s)\lambda_i^\alpha}\right.\right.\right.\right.$$

$$\left.\left.\left.\left. - e^{-(t_m-u)\lambda_i^\alpha}\right)(u-s)^{H-\frac{3}{2}}e^{-\mu(u-s)}\mathrm{d}u\sigma_i\phi_i(x)\mathrm{d}\beta^i(s)\right\|^2\right]\right)^{\frac{1}{2}}$$

$$+ \sum_{k=0}^{t_m/\widetilde{\tau}-1} \left(\mathrm{E}\left[\left\|\sum_{i=1}^{N} \int_{k\widetilde{\tau}}^{(k+1)\widetilde{\tau}} \int_s^{(k+1)\widetilde{\tau}} \left(e^{-(t_m-u)\lambda_i^\alpha}\right.\right.\right.\right.$$

$$\left.\left.\left.\left. - e^{-(t_m-k\widetilde{\tau})\lambda_i^\alpha}\right)(u-s)^{H-\frac{1}{2}}e^{-\mu(u-s)}\mathrm{d}u\sigma_i\mu\phi_i(x)\mathrm{d}\beta^i(s)\right\|^2\right]\right)^{\frac{1}{2}}$$

$$= I + II + III.$$

(3.34)

Let $0 \leq \theta < 2 \cdot \min\{H, 1\}$ and $0 < \theta_1 \leq 1$. Itô's isometry leads to

$$I \lesssim \tilde{\tau}^{\frac{1-\theta}{2}} \sum_{k=1}^{t_m/\tilde{\tau}-1} \left(\sum_{i=1}^{N} \left(\int_{(k-1)\tilde{\tau}}^{(k+1)\tilde{\tau}} (t_m - r)^{-1+\theta_1} dr \right)^2 \lambda_i^{2\alpha\theta_1-\alpha\theta-2\rho} \right)^{\frac{1}{2}}$$

$$\lesssim \tilde{\tau}^{\frac{1-\theta}{2}} \sum_{k=1}^{t_m/\tilde{\tau}-1} \left(\int_{k\tilde{\tau}}^{(k+1)\tilde{\tau}} (t_m - r)^{-1+\theta_1} dr + \int_{(k-1)\tilde{\tau}}^{k\tilde{\tau}} (t_m - r)^{-1+\theta_1} dr \right)$$

$$\times \left(\sum_{i=1}^{N} \lambda_i^{2\alpha\theta_1-\alpha\theta-2\rho} \right)^{\frac{1}{2}}$$

$$\lesssim \frac{\tilde{\tau}^{\frac{1-\theta}{2}}}{\theta_1} \left(\sum_{i=1}^{N} \lambda_i^{2\alpha\theta_1-\alpha\theta-2\rho} ds \right)^{\frac{1}{2}}. \tag{3.35}$$

For the third and fourth inequality, we use the fact that $e^{-ay} \lesssim |ay|^{-b}$ $(a, b, y > 0)$. Similarly, for II, let $0 \leq \theta = 2\eta - 2\delta < 2H$ and $\delta > \frac{1}{2}$ H. We have

$$II \lesssim \left(\sum_{i=1}^{N} \lambda_i^{-2\rho} \int_{t_m-\tilde{\tau}}^{t_m} \lambda_i^{-2\alpha(\eta-\delta)} (t_m - s)^{2H-2\eta+2\delta-1} ds \right)^{\frac{1}{2}}$$

$$+ \sum_{k=0}^{t_m/\tilde{\tau}-2} \left(\sum_{i=1}^{N} \int_{k\tilde{\tau}}^{(k+1)\tilde{\tau}} \lambda_i^{-2\rho} e^{-(t_m-s)\lambda_i^{\alpha}} \left(e^{-\frac{t_m-(k+2)\tilde{\tau}}{2}\lambda_i^{\alpha}} \right. \right.$$

$$\left. \left. - e^{-\frac{t_m-k\tilde{\tau}}{2}\lambda_i^{\alpha}} \right) \left(\int_s^{(k+1)\tilde{\tau}} \lambda_i^{\frac{\alpha(1-\theta)}{2}} (u - s)^{H-1-\frac{\theta}{2}} du \right)^2 ds \right)^{\frac{1}{2}}$$

$$\lesssim \frac{1}{2H-\theta} \sum_{k=0}^{t_m/\tilde{\tau}-2} \int_{k\tilde{\tau}}^{(k+2)\tilde{\tau}} (t_m - r)^{-1+\theta_1} dr \left(\sum_{i=1}^{N} \lambda_i^{2\alpha\theta_1-2\rho-\alpha\theta} \tilde{\tau}^{2H-\theta} \right)^{\frac{1}{2}}$$

$$\lesssim \frac{\tilde{\tau}^{H-\frac{\theta}{2}}}{\theta_1(2H-\theta)} \left(\sum_{i=1}^{N} \lambda_i^{2\alpha\theta_1-2\rho-\alpha\theta} \right)^{\frac{1}{2}}. \tag{3.36}$$

Similar to I, we have

$$
III \lesssim \sum_{k=1}^{t_m/\widetilde{\tau}-1} \int_{(k-1)\widetilde{\tau}}^{(k+1)\widetilde{\tau}} (t_m - r)^{-1+\theta_1} dr \left(\sum_{i=1}^{N} \lambda_i^{2\alpha\theta_1 - 2\rho} \right.
$$

$$
\times \int_{k\widetilde{\tau}}^{(k+1)\widetilde{\tau}} \left(\int_s^{(k+1)\widetilde{\tau}} e^{-\frac{t_m-u}{2}\lambda_i^\alpha} (u-s)^{H-\frac{1}{2}} du \right)^2 ds \right)^{\frac{1}{2}}
$$

$$
+ \left(\sum_{i=1}^{N} \lambda_i^{-2\rho} \int_0^{\widetilde{\tau}} \left(\int_s^{\widetilde{\tau}} e^{-\frac{t_m-u}{2}\lambda_i^\alpha} (u-s)^{H-\frac{1}{2}} du \right)^2 ds \right)^{\frac{1}{2}}
$$

$$
\lesssim \sum_{k=0}^{t_m/\widetilde{\tau}-1} \int_{(k-1)\widetilde{\tau}}^{(k+1)\widetilde{\tau}} (t_m - r)^{-1+\theta_1} dr
$$

$$
\times \left(\sum_{i=1}^{N} \lambda_i^{2\alpha\theta_1 - \alpha\theta - 2\rho} \int_{k\widetilde{\tau}}^{(k+1)\widetilde{\tau}} ((k+1)\widetilde{\tau} - s)^{2H+1-\theta} ds \right)^{\frac{1}{2}}
$$

$$
\lesssim \frac{\widetilde{\tau}^{H+1-\frac{\theta}{2}}}{\theta_1} \left(\sum_{i=1}^{N} \lambda_i^{2\alpha\theta_1 - 2\rho - \alpha\theta} \right)^{\frac{1}{2}}. \tag{3.37}
$$

While $\rho > \frac{1}{2}$, we choose $\theta = 0$, $\theta_1 = \min\left\{\frac{\rho-\frac{1}{2}}{2\alpha}, 1\right\}$; for $\frac{1}{2} - \frac{2-\epsilon}{2}\alpha \cdot \min\{H,1\} < \rho \leq \frac{1}{2}$, let $\theta = \frac{1-2\rho+\epsilon\alpha\cdot\min\{H,1\}}{\alpha}$, $\theta_1 = \frac{\epsilon\cdot\min\{H,1\}}{4}$. Then combining Eqs. (3.35), (3.36) and (3.37) results in

$$
I + II + III
$$

$$
\lesssim \begin{cases} (2\rho-1)^{-\frac{1}{2}}\frac{1}{H(2\rho-1)}\widetilde{\tau}^H, & \rho > \frac{1}{2}, \\ (\epsilon\alpha H\gamma)^{-\frac{1}{2}}\frac{1}{\epsilon}\widetilde{\tau}^{H-\frac{1-2\rho+\epsilon\alpha\cdot\min\{H,1\}}{2\alpha}}, \\ \quad \frac{1}{2} - \frac{2-\epsilon}{2}\alpha\cdot\min\{H,1\} < \rho \leq \frac{1}{2}. \end{cases}
$$

As $H > \frac{1}{2}$, similar to the derivation of Eq. (3.37), using Eqs. (3.15) and (3.16) leads to

$$
\left\| \sum_{k=0}^{t_m/\widetilde{\tau}-1} \int_{k\widetilde{\tau}}^{(k+1)\widetilde{\tau}} (S_N(t_m-s) - S_N(t_m-k\widetilde{\tau})) \, dB_{H,\mu}(s) \right\|_{L^2(D,U)}
$$

$$
\lesssim \begin{cases} (2\rho-1)^{-\frac{1}{2}}\frac{1}{H(2\rho-1)}\widetilde{\tau}^H, & \rho > \frac{1}{2}, \\ (\epsilon\alpha H\gamma)^{-\frac{1}{2}}\frac{1}{\epsilon}\widetilde{\tau}^{H-\frac{1-2\rho+\epsilon\alpha\cdot\min\{H,1\}}{2\alpha}}, \\ \quad \frac{1}{2} - \frac{2-\epsilon}{2}\alpha\cdot\min\{H,1\} < \rho \leq \frac{1}{2}. \end{cases}
$$

$\square$

The following theorem shows the convergence rates of time discretization.

Theorem 3.4 *Let $u^N(t)$ be the mild solution of Eq. (3.21). Suppose that Assumptions 1-2 are satisfied, $\|u(0)\|_{L^2(D,\dot{U}^{\gamma+\alpha})} < \infty$, $0 < \tau < 1$ and $\epsilon = \frac{1}{|\log \tau|}$. Then*

(i) *For $\rho > \frac{1}{2}$,*

$$\left\|u^N(t_m) - u_m^N\right\|_{L^2(D,U)}$$
$$\lesssim \tau^{\min\{H,1\}} \left(\frac{1}{\min\{\gamma, \alpha H, 2\rho - 1\}} + \|u_0\|_{L^2(D,\dot{U}^\gamma)} \right);$$

(ii) *For $\frac{1}{2} - \frac{(2-\epsilon)}{2}\alpha \cdot \min\{H,1\} < \rho \le \frac{1}{2}$,*

$$\left\|u^N(t_m) - u_m^N\right\|_{L^2(D,U)} \lesssim \tau^{\frac{2\rho-1+2\alpha \cdot \min\{H,1\}}{2\alpha}}$$
$$\left(\frac{|\log \tau|}{\min\{H,1\}} + \|u_0\|_{L^2(D,\dot{U}^\gamma)} \right).$$

Proof: Combining Eq. (3.33) and Proposition 3.2, there exists

$$\left\|u^N(t_m) - u_m^N\right\|_{L^2(D,U)}$$
$$\lesssim \left\|z^N(t_m) - z_m^N\right\|_{L^2(D,U)}$$
$$+ \left\| \sum_{k=0}^{t_m/\tilde{\tau}-1} \int_{k\tilde{\tau}}^{(k+1)\tilde{\tau}} (S_N(t_m - s) - S_N(t_m - k\tilde{\tau}))\, \mathrm{d}B_{H,\mu}(s) \right\|_{L^2(D,U)}.$$

Thus, we just need to estimate the bound of $\|z^N(t_m) - z_m^N\|_{L^2(D,U)}$. Let $e_m = z^N(t_m) - z_m^N$ and $\chi \in U$. From Eqs. (3.25) and (3.32), we have

$$\langle e_{m+1} - e_m, \chi \rangle = -\int_{t_m}^{t_{m+1}} \left\langle \left(A_N^\alpha z^N(s) - A_N^\alpha z_{m+1}^N \right), \chi \right\rangle \mathrm{d}s$$
$$+ \int_{t_m}^{t_{m+1}} \left\langle \left(f_N\left(u^N(s)\right) - f_N\left(u_m^N\right) \right), \chi \right\rangle \mathrm{d}s.$$

Set $\chi = e_{m+1}$. Using the fact $(a - b)a = \frac{1}{2}(a^2 - b^2) + \frac{1}{2}(a - b)^2$ in the left-hand side of the above equation, we get

$$
\frac{1}{2}\left(\mathrm{E}\left[\|e_{m+1}\|^2\right] - \mathrm{E}\left[\|e_m\|^2\right]\right) + \frac{1}{2}\mathrm{E}\left[\|e_{m+1} - e_m\|^2\right]
$$
$$
= \mathrm{E}\left[-\int_{t_m}^{t_{m+1}}\left\langle\left(A_N^\alpha z^N(s) - A_N^\alpha z_{m+1}^N\right), e_{m+1}\right\rangle ds\right]
$$
$$
+ \mathrm{E}\left[\int_{t_m}^{t_{m+1}}\left\langle f_N\left(u^N(s)\right) - f_N\left(u_m^N\right), e_{m+1}\right\rangle ds\right]. \quad (3.38)
$$

To obtain the estimate of $\mathrm{E}\left[\|e_m\|^2\right]$, we need to bound the right-hand side of Eq. (3.38).

As $\rho > \frac{1}{2}$, from Theorem 3.3, it has

$$
\mathrm{E}\left[-\int_{t_m}^{t_{m+1}}\left\langle\left(A_N^\alpha z^N(s) - A_N^\alpha z_{m+1}^N\right), e_{m+1}\right\rangle ds\right]
$$
$$
= -\mathrm{E}\left[\int_{t_m}^{t_{m+1}}\left\langle A_N^{\frac{\alpha}{2}}\left(z^N(s) - z_{m+1}^N\right), A_N^{\frac{\alpha}{2}}e_{m+1}\right\rangle ds\right]
$$
$$
\lesssim \frac{1}{2}\int_{t_m}^{t_{m+1}}\mathrm{E}\left[\left\|A_N^{\frac{\alpha}{2}}\left(z^N(s) - z^N(t_{m+1})\right)\right\|^2\right] ds
$$
$$
\lesssim \tau^{\min\{\frac{\gamma}{\alpha}+1, 3\}}\left(\frac{1}{\min\{\gamma\alpha H, \gamma^2\}} + \mathrm{E}\left[\left\|A^{\frac{\gamma+\alpha}{2}}u_0\right\|^2\right]\right). \quad (3.39)
$$

In the first and second inequalities, we use the Hölder inequality and Young's inequality, respectively. Combining the Hölder inequality, Corollary 3.1, and Proposition 3.2 leads to

$$
\mathrm{E}\left[\int_{t_m}^{t_{m+1}}\left\langle f_N\left(u^N(s)\right) - f_N\left(u_m^N\right), e_{m+1}\right\rangle ds\right]
$$
$$
= \mathrm{E}\left[\int_{t_m}^{t_{m+1}}\left\langle f_N\left(u^N(s)\right) - f_N\left(u^N(t_m)\right), e_{m+1}\right\rangle ds\right]
$$
$$
+ \mathrm{E}\left[\int_{t_m}^{t_{m+1}}\left\langle f_N\left(u^N(t_m)\right) - f_N\left(u_m^N\right), e_{m+1}\right\rangle ds\right]
$$
$$
\lesssim \tau^{1+\min\{2H, 2\}}\left(\frac{1}{\min\{\gamma\alpha H, \alpha H^2, 2\rho H - H\}} + \mathrm{E}\left[\left\|A^{\frac{\gamma}{2}}u_0\right\|^2\right]\right)
$$
$$
+ \tau\mathrm{E}\left[\|e_{m+1}\|^2 + \|e_m\|^2\right]. \quad (3.40)
$$

Combining Eqs. (3.38), (3.39), and (3.40), we have

$$\frac{1}{2} \left(\mathrm{E}\left[\|e_{m+1}\|^2\right] - \mathrm{E}\left[\|e_m\|^2\right] \right)$$

$$\lesssim \tau^{1+\min\{2H,2\}} \left(\frac{1}{\min\{\gamma\alpha H, \alpha H^2, 2\rho H - H\}} + \mathrm{E}\left[\left\|A^{\frac{\gamma+\alpha}{2}} u_0\right\|^2\right] \right)$$

$$+ \tau \mathrm{E}\left[\|e_{m+1}\|^2 + \|e_m\|^2\right]. \tag{3.41}$$

Summing m in Eqs. (3.41) from 0 to $\tilde{m}$ ($0 \leq \tilde{m} \leq M - 1$) gives

$$\mathrm{E}\left[\left\|e_{\tilde{m}+1}\right\|^2\right] \lesssim \tau^{\min\{2H,2\}}$$

$$\times \left(\frac{1}{\min\{\alpha H^2, 2\rho H - H\}} + \mathrm{E}\left[\left\|A^{\frac{\gamma+\alpha}{2}} u_0\right\|^2\right] \right) + \sum_{m=0}^{\tilde{m}} \tau \mathrm{E}\left[\|e_{m+1}\|^2\right].$$

By using the discrete Grönwall inequality, we have

$$\mathrm{E}\left[\|e_{m+1}\|^2\right] \lesssim \tau^{\min\{2H,2\}}$$

$$\times \left(\frac{1}{\min\{\alpha H^2, 2\rho H - H\}} + \mathrm{E}\left[\left\|A^{\frac{\gamma+\alpha}{2}} u_0\right\|^2\right] \right).$$

The case (ii) can be similarly proved. □

Combining Theorems 3.2 and 3.4, we get the error bounds for the full discretization.

Theorem 3.5 *Let Assumption 1-2 be satisfied and $u(t)$ be the mild solution of Eq. (3.4). If $\|u(0)\|_{L^2(D,\dot{U}^{\gamma+\alpha})} < \infty$, and $0 < \tau < 1$, then we have*
(i) *For $\rho > \frac{1}{2}$,*

$$\left\|u(t_m) - u_m^N\right\|_{L^2(D,U)}$$

$$\lesssim \lambda_{N+1}^{-\frac{2\rho-1+2\alpha\cdot\min\{H,1\}}{2}} \left(\frac{\alpha^{-\frac{1}{2}} \log \lambda_{N+1}}{\min\{H,1\}} + \|u_0\|_{L^2(D,\dot{U}^\gamma)} \right)$$

$$+ \tau^{\min\{H,1\}} \left(\frac{1}{\min\{\gamma, \alpha H, 2\rho - 1\}} + \|u_0\|_{L^2(D,\dot{U}^\gamma)} \right);$$

(ii) *For* $\frac{1}{2} - \frac{(2-\epsilon)}{2}\alpha \cdot \min\{H, 1\} < \rho \leq \frac{1}{2}$,

$$\left\| u(t_m) - u_m^N \right\|_{L^2(D,U)}$$

$$\lesssim \lambda_{N+1}^{-\frac{2\rho-1+2\alpha\cdot\min\{H,1\}}{2}} \left(\frac{\alpha^{-\frac{1}{2}}\log\lambda_{N+1}}{\min\{H, 1\}} + \|u_0\|_{L^2(D,\dot{U}^\gamma)} \right)$$

$$+ \tau^{\frac{2\rho-1+2\alpha\cdot\min\{H,1\}}{2\alpha}} \left(\frac{|\log\tau|}{\min\{H, 1\}} + \|u_0\|_{L^2(D,\dot{U}^\gamma)} \right).$$

3.2.1.4 *Numerical experiments*

We present the simulation results to show the convergence behavior of the full discretization scheme and the effect of the parameters H, α, and ρ on the convergence rates. All numerical errors are given in the sense of mean-squared L^2-norm.

We solve (3.3) in the two-dimensional domain $D = (0, 1) \times (0, 1)$ by the proposed method with Dirichlet eigenpairs $\lambda_{i,j} = \pi^2 (i^2 + j^2)$, $\phi_{i,j} = 2\sin(i\pi x_1)\sin(j\pi x_2)$, $x = (x_1, x_2)$, and $i, j = 1, 2, \ldots, N_0$ ($N = N_0^2$). The numerical results with a smooth initial data $u(x, 0) = x_1^2 x_2^2$ and $f(u(x, t)) = u(x, t)$ are presented in Tables 3.1–3.5, where u_m^N denotes the numerical solution with fixed time step size $\tau = \frac{T}{M}$ at time $t = m\tau$. Since the exact solutions of Eq. (3.3) are unknown, we use the following formulas to calculate the convergence rates:

convergence rate in space

$$= \frac{\ln \left(\left\| u_M^N - u_M^{N/2.25} \right\|_{L^2(D,U)} / \left\| u_M^{2.25N} - u_M^N \right\|_{L^2(D,U)} \right)}{\ln 2.25},$$

convergence rate in time

$$= \frac{\ln \left(\left\| u_M^N - u_{M/1.5}^N \right\|_{L^2(D,U)} / \left\| u_{1.5M}^N - u_M^N \right\|_{L^2(D,U)} \right)}{\ln 1.5}.$$

In the numerical simulations, the errors $\left\| u_M^{2.25N,M} - u_M^{N,M} \right\|_{L^2(D,U)}$ are calculated by Monte Carlo method, i.e.,

$$\left\| u_M^{2.25N,M} - u_M^{N,M} \right\|_{L^2(D,U)} \approx \left(\frac{1}{K} \sum_{k=1}^{K} \left\| u_{M,k}^{2.25N,M} - u_{M,k}^{N,M} \right\|^2 \right)^{\frac{1}{2}}.$$

We take $K = 1000$ as the number of the simulation trajectories. The symbol k represents the k-th trajectory.

Table 3.1 Time convergence rates with $N = 50$, $T = 0.5$, $\alpha = 0.5$, $\rho = 0.75$, and $\lambda = 1$

M	$H = 0.4$	Rate	$H = 0.8$	Rate	$H = 1.2$	Rate	$H = 1.6$	Rate
32	8.895e-02		8.241e-03		1.775e-03		9.924e-04	
48	7.578e-02	0.395	5.976e-03	0.793	1.206e-03	0.952	6.540e-04	1.028
72	6.420e-02	0.409	4.314e-03	0.804	8.118e-04	0.977	4.326e-04	1.019

From Table 3.1, one can see that the time convergence rates increase with the increase of H, when $\rho > \frac{1}{2}$. When $0 < H < 1$, the proposed methods have H-order convergence in time. As $H \geq 1$, the convergence rate is first-order in time. The numerical results confirm the error estimate in Theorem 3.5. Table 3.2 demonstrates

Table 3.2 Time convergence rates with $N = 50$, $T = 0.5$, $\alpha = 0.8$, $\rho = 0.4$, and $\lambda = 0.5$

M	$H = 0.6$	Rate	$H = 0.8$	Rate	$H = 1$	Rate	$H = 1.2$	Rate
32	6.407e-02		1.996e-02		6.811e-03		2.711e-03	
48	5.413e-02	0.416	1.553e-02	0.619	5.024e-03	0.751	1.902e-03	0.874
72	4.525e-02	0.442	1.200e-02	0.637	3.625e-03	0.805	1.324e-03	0.893

that the time convergence rates increase with the increase of H, when $0 < \rho \leq \frac{1}{2}$. The theoretical convergence rates are near to $\frac{2\rho - 1 + 2\alpha \cdot \min\{H,1\}}{2\alpha}$ in time. The numerical simulation agrees well with the theoretical results.

Table 3.3 Space convergence rates with $M = 2000$, $T = 0.25$, $\alpha = 0.3$, $\rho = 0.75$, and $\lambda = 0.5$

N	$H = 0.35$	Rate	$H = 0.7$	Rate	$H = 1.05$	Rate	$H = 1.4$	Rate
256	4.190e-02		1.431e-02		8.352e-03		6.969e-03	
576	3.195e-02	0.334	1.009e-02	0.431	5.525e-03	0.510	4.515e-03	0.536
1296	2.402e-02	0.352	6.967e-03	0.457	3.619e-03	0.522	2.924e-03	0.536

Table 3.3 shows that the space convergence rates increase with the increase of H as $H \leq 1$; and the convergence rates tend to 0.55 as $H > 1$. The numerical results confirm the theoretical prediction

of convergence rates close to $\alpha \cdot \min\{H, 1\} + \rho - \frac{1}{2}$, given in Theorem 3.5.

Table 3.4 Space convergence rates with $M = 2000$, $T = 0.25$, $H = 0.8$, $\rho = 0.75$ and $\lambda = 0.5$

N	$\alpha = 0.2$	Rate	$\alpha = 0.4$	Rate	$\alpha = 0.6$	Rate	$\alpha = 0.8$	Rate
256	2.450e-02		5.997e-03		1.637e-03		4.306e-04	
576	1.793e-02	0.385	3.938e-03	0.519	9.389e-04	0.686	2.106e-04	0.882
1296	1.306e-02	0.391	2.532e-03	0.545	5.287e-04	0.708	1.021e-04	0.893

As $\alpha = 0.2$, 0.4, 0.6, and 0.8, the theoretical convergence rates in space are near to 0.41, 0.57, 0.73, and 0.89, respectively. The numerical results obey the theoretical prediction and show that the space convergence rates increase with the increase of α in Table 3.4.

Table 3.5 Space convergence rates with $M = 4000$, $T = 0.25$, $H = 1.2$, $\alpha = 0.5$, and $\lambda = 0.5$

N	$\rho = 0.75$	Rate	$\rho = 1.25$	Rate	$\rho = 1.75$	Rate	$\rho = 2.25$	Rate
256	1.446e-03		5.964e-05		2.665e-06		1.279e-07	
576	8.185e-04	0.702	2.284e-05	1.184	6.870e-07	1.672	2.173e-08	2.186
1296	4.581e-04	0.716	8.649e-06	1.198	1.762e-07	1.678	3.635e-09	2.205

Table 3.5 shows that the space convergence rates increase with the increase of ρ. That is, the space convergence rates of the spectral Galerkin method can be continuously enhanced by improving the regularity of the mild solution in space.

3.3 NUMERICAL SCHEMES FOR STOCHASTIC FRACTIONAL WAVE EQUATION

3.3.1 Higher Order Approximation for Stochastic Space Fractional Wave Equation

The wave propagation in ideal medium is well described by the classical wave equation $\partial^2 u(x,t)/\partial t^2 = \Delta u(x,t)$. However, sometimes the classical wave equation fails to model the wave propagations in complex inhomogeneous media (e.g., viscous damping in the seismic isolation of buildings, medical ultrasound, and seismic wave propagation [35, 134, 171]), because of their power-law attenuations. One of the most effective ways to characterize the wave propagation with power-law attenuations is to resort to the nonlocal operator—the infinitesimal generator (fractional Laplacian) of a process obtained by subordinating a killed Brownian motion.

Moreover, we are also concerned with the external noises that possibly affect the wave propagation. Two most popular external noises are white noise and fractional Gaussian noise, both of which are considered in this section. The fractional Gaussian noise is defined as the formal derivative of the fractional Brownian motion (fBm) $\beta_H(t)$, which is Gaussian process with an index $H \in (0, 1)$. The fBm has two unique properties: self-similarity and stationary increments [58,59]. As $H = \frac{1}{2}$, the fBm reduces to a standard Brownian motion. The formal derivative of Brownian motion is white noise. For $H \neq \frac{1}{2}$, unlike Brownian motion, the fBm exhibits long-range dependence: the behavior of the process after a given time t depends on the situation at t and the whole history of the process up to time t [45]. According to the properties of the fBm and Brownian motion, one can choose the appropriate noise in practical applications.

With the above introduction of nonlocal operator and the external noise, the model we discuss is the stochastic wave equation

$$\begin{cases} \frac{d\dot{u}(x,t)}{dt} = -(-\Delta)^\alpha u(x,t) + f\left(u(x,t)\right) + \dot{B}_H(x,t) & \text{in } D \times (0,T], \\ u(x,0) = u_0, \ \dot{u}(x,0) = v_0 & \text{in } D, \\ u(x,t) = 0, & \text{in } \partial D \times (0,T], \end{cases}$$

$$(3.42)$$

where $\dot{u}(x,t)$ is the first-order time derivative of $u(x,t)$, d/dt means the partial derivative with respect to t, f is the source term, $D \subset \mathbb{R}^d$ $(d = 1, 2, 3)$, and $\dot{B}_H(x,t)$ is the formal derivative of the infinite dimensional space-time Gaussian process $B_H(x,t)$ with $0 < \alpha \leq 1$ and $\frac{1}{2} \leq H < 1$. The definition of infinite dimensional fBm $B_H(x,t)$ is given in Assumption 2.

3.3.1.1 Regularity of the solution

To begin with we can give a system of equations by coupling (3.42) and $du(x,t) = \dot{u}(x,t)dt$. The system of equations is beneficial to analyze the regularity of the mild solution of (3.42), including existence, uniqueness, and time Hölder continuity. Moreover, the system of equations is transformed into an equivalent form, which will be used to obtain the approximation of (3.42).

In the interest of brevity and readability, we use the following equation instead of (3.42)

$$\begin{cases} d\dot{u}(t) = -A^\alpha u(t)dt + f\left(u(t)\right)dt + dB_H(t), & \text{in } D \times (0,T], \\ u(0) = u_0, \dot{u}(0) = v_0 & \text{in } D, \\ u(t) = 0, & \text{in } \partial D, \end{cases}$$

$$(3.43)$$

where $u(t) = u(x,t)$ and $B_H(t) = B_H(x,t)$. Let $v(t) = \dot{u}(t)$. Then

$$dX(t) = \Lambda X(t)dt + \begin{bmatrix} 0 \\ f(u(t)) \end{bmatrix} dt + \begin{bmatrix} 0 \\ I \end{bmatrix} dB_H(t), \qquad (3.44)$$

where

$$X(t) = \begin{bmatrix} u(t) \\ v(t) \end{bmatrix}, \qquad \Lambda = \begin{bmatrix} 0 & I \\ -A^\alpha & 0 \end{bmatrix}.$$

Then a formal mild solution $X(t)$ for (3.44) is given as

$$X(t) = e^{\Lambda t} X(0) + \int_0^t e^{\Lambda(t-s)} \begin{bmatrix} 0 \\ f\left(u(s)\right) \end{bmatrix} ds$$

$$+ \int_0^t e^{\Lambda(t-s)} \begin{bmatrix} 0 \\ I \end{bmatrix} dB_H(s),$$

$$(3.45)$$

where $e^{\Lambda t}$ can be expressed as

$$e^{\Lambda t} = \begin{bmatrix} \cos\left(A^{\frac{\alpha}{2}}t\right) & A^{-\frac{\alpha}{2}}\sin\left(A^{\frac{\alpha}{2}}t\right) \\ -A^{\frac{\alpha}{2}}\sin\left(A^{\frac{\alpha}{2}}t\right) & \cos\left(A^{\frac{\alpha}{2}}t\right) \end{bmatrix}. \qquad (3.46)$$

The definitions of cosine operator $\cos\left(A^{\frac{\alpha}{2}}t\right)$ and sine operator $\sin\left(A^{\frac{\alpha}{2}}t\right)$ are given as

$$\begin{aligned} \sin\left(A^\alpha t\right)u(t) &= \sum_{i=1}^\infty \sin\left(\lambda_i^\alpha t\right)\langle u(t), \phi_i(x)\rangle \phi_i(x) \\ &= \sum_{i=1}^\infty \sum_{j=1}^\infty (-1)^{j-1}\frac{(\lambda_i^\alpha t)^{2j-1}}{(2j-1)!}\langle u(t), \phi_i(x)\rangle \phi_i(x) \end{aligned}$$

and

$$\begin{aligned} \cos\left(A^\alpha t\right)u(t) &= \sum_{i=1}^\infty \cos\left(\lambda_i^\alpha t\right)\langle u(t), \phi_i(x)\rangle \phi_i(x) \\ &= \sum_{i=1}^\infty \sum_{j=0}^\infty (-1)^j\frac{(\lambda_i^\alpha t)^{2j}}{(2j)!}\langle u(t), \phi_i(x)\rangle \phi_i(x). \end{aligned}$$

Substituting (3.46) into (3.45), then two components of $X(t)$ are obtained as

$$u(t) = \cos\left(A^{\frac{\alpha}{2}}t\right)u_0 + A^{-\frac{\alpha}{2}}\sin\left(A^{\frac{\alpha}{2}}t\right)v_0$$

$$+ \int_0^t A^{-\frac{\alpha}{2}}\sin\left(A^{\frac{\alpha}{2}}(t-s)\right)f\left(u(s)\right)\mathrm{d}s$$

$$+ \int_0^t A^{-\frac{\alpha}{2}}\sin\left(A^{\frac{\alpha}{2}}(t-s)\right)\mathrm{d}B_H(s),$$

$$v(t) = -A^{\frac{\alpha}{2}}\sin\left(A^{\frac{\alpha}{2}}t\right)u_0 + \cos\left(A^{\frac{\alpha}{2}}t\right)v_0 \tag{3.47}$$

$$+ \int_0^t \cos\left(A^{\frac{\alpha}{2}}(t-s)\right)f\left(u(s)\right)\mathrm{d}s$$

$$+ \int_0^t \cos\left(A^{\frac{\alpha}{2}}(t-s)\right)\mathrm{d}B_H(s).$$

In order to obtain the regularity of $u(t)$ and $v(t)$, we need to consider the regularity estimate of the stochastic integral in (3.47). For $\frac{1}{2} \le H < 1$ and $p > 1$, the Burkhölder-Davis-Gundy inequality [151] implies

$$\mathbb{E}\left[\left\|\int_0^t A^{\frac{\gamma-\alpha}{2}}\sin\left(A^{\frac{\alpha}{2}}(t-s)\right)\mathrm{d}B(s)\right\|^p\right] \tag{3.48}$$

$$\le C_p\left(\int_0^t \sum_{i=1}^{\infty}\left|\lambda_i^{\frac{\gamma-\alpha-2\rho}{2}}\sin\left(\lambda_i^{\frac{\alpha}{2}}(t-s)\right)\right|^2\mathrm{d}s\right)^{\frac{p}{2}}$$

$$\le C_p\left(t\sum_{i=1}^{\infty}i^{\frac{2(\gamma-\alpha-2\rho)}{d}}\right)^{\frac{p}{2}}$$

and

$$\mathbb{E}\left[\left\|\int_0^t A^{\frac{\gamma-\alpha}{2}}\sin\left(A^{\frac{\alpha}{2}}(t-s)\right)\mathrm{d}B_H(s)\right\|^p\right]$$

$$\le C_H^p C_p\left(\int_0^t \sum_{i=1}^{\infty}\lambda_i^{\gamma-\alpha-2\rho}\left(\int_s^t \left(\frac{s}{r}\right)^{\frac{1}{2}-H}(r-s)^{H-\frac{3}{2}}\mathrm{d}r\right)^2\mathrm{d}s\right)^{\frac{p}{2}}$$

$$\le C_H^p C_p\left(\int_0^t \sum_{i=1}^{\infty}\lambda_i^{\gamma-\alpha-2\rho}\left(\frac{s}{t}\right)^{1-2H}\left(\int_s^t (r-s)^{H-\frac{3}{2}}\mathrm{d}r\right)^2\mathrm{d}s\right)^{\frac{p}{2}}$$

$$\le C_H^p C_p\left(t^{2H}\sum_{i=1}^{\infty}i^{\frac{2(\gamma-\alpha-2\rho)}{d}}\right)^{\frac{p}{2}}.$$

$$\tag{3.49}$$

When $\frac{2(\gamma-\alpha-2\rho)}{d} < -1$, then the infinite series $\sum\limits_{i=1}^{\infty} i^{\frac{2(\gamma-\alpha-2\rho)}{d}} < \infty$.

One can obtain the following regularity results of the mild solution $u(t)$ and $v(t)$ by using the above estimates, Lemma 3.1, and (3.47).

Theorem 3.6 *Suppose that Assumptions 1 and 2 are satisfied,* $\|u_0\|_{L^p(D,\dot{U}^\gamma)} < \infty$, $\|v_0\|_{L^p(D,\dot{U}^{\gamma-\alpha})} < \infty$, $\varepsilon > 0$, $\gamma = \alpha + 2\rho - \frac{d+\varepsilon}{2}$, *and* $\gamma > 0$. *Then there exists a unique mild solution* $X(t)$ *for* (3.44) *and*

$$\|u(t)\|_{L^p(D,\dot{U}^\gamma)} + \|v(t)\|_{L^p(D,\dot{U}^{\gamma-\alpha})}$$

$$\lesssim \frac{t^H}{\varepsilon} + \|u_0\|_{L^p(D,\dot{U}^\gamma)} + \|v_0\|_{L^p(D,\dot{U}^{\gamma-\alpha})}. \tag{3.50}$$

Furthermore,

   (i) *for* $\gamma \leq \alpha$,

$$\|u(t) - u(s)\|_{L^2(D,U)} \lesssim (t-s)^{\frac{\gamma}{\alpha}}$$

$$\times \left(\frac{t^H}{\varepsilon} + \|u_0\|_{L^2(D,\dot{U}^\gamma)} + \|v_0\|_{L^2(D,\dot{U}^{\gamma-\alpha})} \right);$$

  (ii) *for* $\gamma > \alpha$,

$$\|u(t) - u(s)\|_{L^2(D,U)} \lesssim (t-s)$$

$$\times \left(t^H + \|u_0\|_{L^2(D,\dot{U}^\gamma)} + \|v_0\|_{L^2(D,\dot{U}^{\gamma-\alpha})} \right).$$

Proof: Let us start with the estimate of $A^{\frac{\gamma}{2}}u(t)$ in $L^p(D,U)$ norm. Combining the triangle inequality, (3.48), (3.49), the expression of $u(t)$ in (3.47), and the assumption of f, we obtain

$$\left\| A^{\frac{\gamma}{2}}u(t) \right\|_{L^p(D,U)}$$

$$\lesssim \left\| A^{\frac{\gamma}{2}} \cos\left(A^{\frac{\alpha}{2}}t\right) u_0 \right\|_{L^p(D,U)} + \left\| A^{\frac{\gamma-\alpha}{2}} \sin\left(A^{\frac{\alpha}{2}}t\right) v_0 \right\|_{L^p(D,U)}$$

$$+ \left\| \int_0^t A^{\frac{\gamma-\alpha}{2}} \sin\left(A^{\frac{\alpha}{2}}(t-s)\right) f(u(s)) \, \mathrm{d}s \right\|_{L^p(D,U)}$$

$$+ \left\| \int_0^t A^{\frac{\gamma-\alpha}{2}} \sin\left(A^{\frac{\alpha}{2}}(t-s)\right) \mathrm{d}B_H(s) \right\|_{L^p(D,U)}$$

$$\lesssim \left\| A^{\frac{\gamma}{2}}u_0 \right\|_{L^p(D,U)} + \left\| A^{\frac{\gamma-\alpha}{2}}v_0 \right\|_{L^p(D,U)}$$

$$+ \int_0^t \left\| A^{\frac{\gamma}{2}}u(s) \right\|_{L^p(D,U)} \, \mathrm{d}s + \frac{t^H}{\varepsilon}.$$

The application of Grönwall's inequality leads to

$$\left\|A^{\frac{\gamma}{2}}u(t)\right\|_{L^p(D,U)} \lesssim \left\|A^{\frac{\gamma}{2}}u_0\right\|_{L^p(D,U)} + \left\|A^{\frac{\gamma-\alpha}{2}}v_0\right\|_{L^p(D,U)} + \frac{t^H}{\varepsilon}.$$

The bound of $\|v(t)\|_{L^p(D,\dot{U}^{\gamma-\alpha})}$ can be achieved in the same way, that is

$$\left\|A^{\frac{\gamma-\alpha}{2}}v(t)\right\|_{L^p(D,U)} \lesssim \left\|A^{\frac{\gamma}{2}}u_0\right\|_{L^p(D,U)} + \left\|A^{\frac{\gamma-\alpha}{2}}v_0\right\|_{L^p(D,U)} + \frac{t^H}{\varepsilon}.$$

Then using above estimates leads to

$$\|u(t)\|_{L^p(D,\dot{U}^{\gamma})} + \|v(t)\|_{L^p(D,\dot{U}^{\gamma-\alpha})}$$
$$\lesssim \frac{t^H}{\varepsilon} + \|u_0\|_{L^p(D,\dot{U}^{\gamma})} + \|v_0\|_{L^p(D,\dot{U}^{\gamma-\alpha})}.$$

Next, we discuss the time Hölder continuity of $u(t)$. Equation (3.47) implies

$$\mathrm{E}\left[\|u(t) - u(s)\|^2\right]$$
$$\lesssim \mathrm{E}\left[\left\|\left(\cos\left(A^{\frac{\alpha}{2}}t\right) - \cos\left(A^{\frac{\alpha}{2}}s\right)\right)u_0\right\|^2\right]$$
$$+ \mathrm{E}\left[\left\|A^{-\frac{\alpha}{2}}\left(\sin\left(A^{\frac{\alpha}{2}}t\right) - \sin\left(A^{\frac{\alpha}{2}}s\right)\right)v_0\right\|^2\right]$$
$$+ \mathrm{E}\left[\left\|\int_s^t A^{-\frac{\alpha}{2}}\sin\left(A^{\frac{\alpha}{2}}(t-r)\right)f(u(r))\,\mathrm{d}r\right\|^2\right]$$
$$+ \mathrm{E}\left[\left\|\int_s^t A^{-\frac{\alpha}{2}}\sin\left(A^{\frac{\alpha}{2}}(t-r)\right)\mathrm{d}B_H(r)\right\|^2\right]$$
$$+ \mathrm{E}\left[\left\|\int_0^s A^{-\frac{\alpha}{2}}\left(\sin\left(A^{\frac{\alpha}{2}}(t-r)\right) - \sin\left(A^{\frac{\alpha}{2}}(s-r)\right)\right)f(u(r))\,\mathrm{d}r\right\|^2\right]$$
$$+ \mathrm{E}\left[\left\|\int_0^s A^{-\frac{\alpha}{2}}\left(\sin\left(A^{\frac{\alpha}{2}}(t-r)\right) - \sin\left(A^{\frac{\alpha}{2}}(s-r)\right)\right)\mathrm{d}B_H(r)\right\|^2\right]$$
$$\lesssim I_1 + I_2 + I_3 + I_4 + I_5 + I_6.$$

For $\frac{1}{2} < H < 1$, the inequality $\cos(a) - \cos(b) \lesssim |a - b|^\theta$ $(0 \le \theta \le 1)$ implies

$$
\begin{aligned}
I_1 &= \mathrm{E}\left[\left\|\sum_i \left(\cos\left(\lambda_i^{\frac{\alpha}{2}} t\right) - \cos\left(\lambda_i^{\frac{\alpha}{2}} s\right)\right) \langle u_0, \phi_i(x) \rangle \phi_i(x)\right\|^2\right] \\
&\lesssim \mathrm{E}\left[\sum_i \lambda_i^\gamma (t - s)^{2 \cdot \min\{\frac{\gamma}{\alpha}, 1\}} \langle u_0, \phi_i(x) \rangle^2\right] \\
&\lesssim (t - s)^{\min\{\frac{2\gamma}{\alpha}, 2\}} \mathrm{E}\left[\left\|A^{\frac{\gamma}{2}} u_0\right\|^2\right].
\end{aligned}
$$

Similar to the derivation of I_1, it holds that

$$
I_2 \lesssim (t - s)^{\min\{\frac{2\gamma}{\alpha}, 2\}} \mathrm{E}\left[\left\|A^{\frac{\gamma - \alpha}{2}} v_0\right\|^2\right].
$$

For the Hölder regularity of the third term, using Equation (3.50) and Assumption 1 leads to

$$
\begin{aligned}
I_3 &\lesssim (t - s) \int_s^t \mathrm{E}\left[\left\|A^{-\frac{\alpha}{2}} \sin\left(A^{\frac{\alpha}{2}}(t - r)\right) f(u(r))\right\|^2\right] \mathrm{d}r \\
&\lesssim (t - s) \int_s^t \mathrm{E}\left[\|u(r)\|^2 + 1\right] \mathrm{d}r \\
&\lesssim (t - s)^2 \left(\mathrm{E}\left[\left\|A^{\frac{\gamma}{2}} u_0\right\|^2\right] + \mathrm{E}\left[\left\|A^{\frac{\gamma - \alpha}{2}} v_0\right\|^2\right] + 1\right).
\end{aligned}
$$

Take $\theta_1 = \min\left\{\frac{\gamma}{2\alpha}, \frac{1}{2}\right\}$. Combining the fact that $|\sin(t)| \lesssim |t|^{\theta_1}$ $(t \in \mathbb{R})$ and Lemma 3.2 leads to

$$
\begin{aligned}
I_4 &= \mathrm{E}\left[\left\|\int_s^t A^{-\frac{\alpha}{2}} \sin\left(A^{\frac{\alpha}{2}}(t - r)\right) \mathrm{d}B_H(r)\right\|^2\right] \\
&\lesssim (t - s)^{2H + \min\{\frac{\gamma}{\alpha}, 1\}} \sum_i \lambda_i^{\frac{\gamma}{2} - \alpha - 2\rho} \\
&\lesssim (t - s)^{2H + \min\{\frac{\gamma}{\alpha}, 1\}}.
\end{aligned}
$$

Using $\sin(a) - \sin(b) \lesssim |a - b|^{\theta_2}$ and the assumption of f with $\theta_2 = \min\left\{\frac{\gamma}{\alpha}, 1\right\}$, we have

$$
\begin{aligned}
I_5 &\lesssim (t - s)^{\min\{\frac{2\gamma}{\alpha}, 2\}} \mathrm{E}\left[\left\|\int_0^s A^{\frac{\gamma - \alpha}{2}} f(u(r)) \mathrm{d}r\right\|^2\right] \\
&\lesssim (t - s)^{\min\{\frac{2\gamma}{\alpha}, 2\}} \left(\mathrm{E}\left[\left\|A^{\frac{\gamma}{2}} u_0\right\|^2\right] + \mathrm{E}\left[\left\|A^{\frac{\gamma - \alpha}{2}} v_0\right\|^2\right] + t^{2H}\right).
\end{aligned}
$$

For $\gamma > \alpha$, by using Lemma 3.2 and the inequality $\sin\left(\lambda_i^{\frac{\alpha}{2}}(t-r)\right) - \sin\left(\lambda_i^{\frac{\alpha}{2}}(s-r)\right) \lesssim \lambda_i^{\frac{\alpha}{2}}(t-s)$, we get the bound of I_6, i.e.,

$$
\begin{aligned}
I_6 &\lesssim \sum_i \lambda_i^{-\alpha-2\rho} \int_0^s \int_0^s \left|\sin\left(\lambda_i^{\frac{\alpha}{2}}(t-r)\right) - \sin\left(\lambda_i^{\frac{\alpha}{2}}(s-r)\right)\right| \\
&\quad \times \left|\sin\left(\lambda_i^{\frac{\alpha}{2}}(t-r_1)\right) - \sin\left(\lambda_i^{\frac{\alpha}{2}}(s-r_1)\right)\right| \times |r-r_1|^{2H-2} dr dr_1 \\
&\lesssim t^{2H}(t-s)^2 \sum_i \lambda_i^{\alpha-\gamma+\gamma-\alpha-2\rho} \\
&\lesssim t^{2H}(t-s)^2.
\end{aligned}
$$

When $\gamma \leq \alpha$, using $\sin\left(\lambda_i^{\frac{\alpha}{2}}(t-r)\right) - \sin\left(\lambda_i^{\frac{\alpha}{2}}(s-r)\right) \lesssim \lambda_i^{\frac{\gamma}{2}}(t-s)^{\frac{\gamma}{\alpha}}$ leads to

$$
\begin{aligned}
I_6 &\lesssim t^{2H}(t-s)^{\frac{2\gamma}{\alpha}} \sum_i \lambda_i^{\gamma-\alpha-2\rho} \\
&\lesssim \frac{l^{2H}}{\varepsilon}(t-s)^{\frac{2\gamma}{\alpha}}.
\end{aligned}
$$

Then collecting the above estimates arrives at

$$
\begin{aligned}
&\mathrm{E}\left[\|u(t) - u(s)\|^2\right] \\
&\lesssim (t-s)^2 \left(t^{2H} + \mathrm{E}\left[\left\|A^{\frac{\gamma}{2}}u_0\right\|^2\right] + \mathrm{E}\left[\left\|A^{\frac{\gamma-\alpha}{2}}v_0\right\|^2\right]\right), \quad \gamma > \alpha
\end{aligned}
$$

and

$$
\begin{aligned}
&\mathrm{E}\left[\|u(t) - u(s)\|^2\right] \\
&\lesssim (t-s)^{\frac{2\gamma}{\alpha}} \left(\frac{t^{2H}}{\varepsilon} + \mathrm{E}\left[\left\|A^{\frac{\gamma}{2}}u_0\right\|^2\right] + \mathrm{E}\left[\left\|A^{\frac{\gamma-\alpha}{2}}v_0\right\|^2\right]\right), \quad \gamma \leq \alpha.
\end{aligned}
$$

When $H = \frac{1}{2}$, the above Hölder regularity results still hold. The proof is completed. $\qquad\square$

In fact, one can get an equivalent form of (3.44) by using variable substitution, the regularity of whose solution is better than the one of the solution of (3.44). Let

$$
Z(t) = X(t) - \int_0^t e^{\Lambda(t-s)} \begin{bmatrix} 0 \\ I \end{bmatrix} dB_H(s), \qquad (3.51)
$$

where

$$Z(t) = \begin{bmatrix} z(t) \\ \dot{z}(t) \end{bmatrix}.$$

If $X(t)$ is the unique mild solution of (3.44), then $Z(t)$ is the unique mild solution of the partial differential equation

$$\frac{\mathrm{d}}{\mathrm{d}t} Z(t) = \Lambda Z(t) + \begin{bmatrix} 0 \\ f(u(t)) \end{bmatrix} \text{ for } t \in (0, T] \text{ with } Z(0) = X(0),$$

(3.52)

where

$$f(u(t)) = f\left(z(t) + \int_0^t A^{-\frac{\alpha}{2}} \sin\left(A^{\frac{\alpha}{2}}(t-s)\right) \mathrm{d}B_H(s)\right).$$

The unique mild solution of (3.52) is given by

$$Z(t) = \mathrm{e}^{\Lambda t} Z(0) + \int_0^t \mathrm{e}^{\Lambda(t-s)} \begin{bmatrix} 0 \\ f(u(t)) \end{bmatrix} \mathrm{d}s. \quad (3.53)$$

Then we can obtain $z(t)$ and $\dot{z}(t)$ as

$$\begin{aligned}
z(t) &= \cos\left(A^{\frac{\alpha}{2}}t\right) u_0 + A^{-\frac{\alpha}{2}} \sin\left(A^{\frac{\alpha}{2}}t\right) v_0 \\
&\quad + \int_0^t A^{-\frac{\alpha}{2}} \sin\left(A^{\frac{\alpha}{2}}(t-s)\right) f(u(s)) \, \mathrm{d}s, \\
\dot{z}(t) &= -A^{\frac{\alpha}{2}} \sin\left(A^{\frac{\alpha}{2}}t\right) u_0 + \cos\left(A^{\frac{\alpha}{2}}t\right) v_0 \\
&\quad + \int_0^t \cos\left(A^{\frac{\alpha}{2}}(t-s)\right) f(u(s)) \, \mathrm{d}s.
\end{aligned}$$

(3.54)

Equation (3.52) will be used to obtain the spatial semi-discretization solution of (3.44). Therefore we give the following estimates, which will be used to discuss the spatial error.

Corollary 3.2 *Suppose that Assumptions 1 and 2 are satisfied,* $\|u_0\|_{L^p(D, \dot{U}^{\gamma+\alpha})} < \infty$, $\|v_0\|_{L^p(D, \dot{U}^\gamma)} < \infty$, $\varepsilon > 0$, $\gamma = \alpha + 2\rho - \frac{d+\varepsilon}{2}$, *and* $\gamma > 0$. *Then there exists a unique mild solution* $Z(t)$ *for* (3.52) *and*

$$\|z(t)\|_{L^p(D, \dot{U}^{\gamma+\alpha})} + \|\dot{z}(t)\|_{L^p(D, \dot{U}^\gamma)} \lesssim \frac{t^H}{\varepsilon} + \|u_0\|_{L^p(D, \dot{U}^{\gamma+\alpha})}$$
$$+ \|v_0\|_{L^p(D, \dot{U}^\gamma)}.$$

The proof of this corollary is very similar to the proof of Theorem 3.6.

3.3.1.2 Galerkin approximation for spatial discretization

The convergence rate of the spectral approximation depends on the regularity of the mild solution in space. Theorem 3.6 and Corollary 3.2 show that $z(t)$ is more regular than $u(t)$ in space; so we obtain the spatial approximation of (3.44) by using the spectral Galerkin method to discretize (3.52) and postprocessing the stochastic integral.

A finite dimensional subspace of U will be needed to implement the Galerkin spatial approximation of (3.52). Denoting the N dimensional subspace of U by U^N, the sequence $\{\phi_1(x), \ldots, \phi_i(x), \ldots, \phi_N(x)\}_{N \in \mathbb{N}}$ is an orthonormal basis of U^N. Then we introduce the projection operator $P_N : U \to U^N$, for $\xi \in U$,

$$P_N \xi = \sum_{i=1}^{N} \langle \xi, \phi_i(x) \rangle \, \phi_i(x)$$

and

$$\langle P_N \xi, \chi \rangle = \langle \xi, \chi \rangle \quad \forall \chi \in U^N. \tag{3.55}$$

To obtain the Galerkin formulation of (3.52), we look for $z^N(t) \in U^N$ and $\dot{z}^N \in U^N$ such that

$$
\begin{bmatrix} \langle \mathrm{d}z^N(t), \chi \rangle \\ \langle \mathrm{d}\dot{z}^N(t), \chi \rangle \end{bmatrix} = \Lambda \begin{bmatrix} \langle z^N(t), \chi \rangle \\ \langle \dot{z}^N(t), \chi \rangle \end{bmatrix} \mathrm{d}t + \begin{bmatrix} 0 \\ \langle f\left(u^N(t)\right), \chi \rangle \end{bmatrix} \mathrm{d}t
$$

$$\tag{3.56}$$

and

$$\left\langle z_0^N, \chi \right\rangle = \langle z_0, \chi \rangle, \quad \left\langle \dot{z}_0^N, \chi \right\rangle = \langle \dot{z}_0, \chi \rangle,$$

where

$$f\left(u^N(t)\right) = f\left(z^N(t) + P_N \int_0^t A^{-\frac{\alpha}{2}} \sin\left(A^{\frac{\alpha}{2}}(t-s)\right) \mathrm{d}B_H(s)\right).$$

The Galerkin formulation of (3.52) is obtained by using (3.55) and (3.56), that is

$$
\begin{bmatrix} \mathrm{d}z^N(t) \\ \mathrm{d}\dot{z}^N(t) \end{bmatrix} = \Lambda \begin{bmatrix} z^N(t) \\ \dot{z}^N(t) \end{bmatrix} \mathrm{d}t + \begin{bmatrix} 0 \\ f_N\left(u^N(t)\right) \end{bmatrix} \mathrm{d}t \tag{3.57}
$$

and
$$z_0^N = P_N u_0, \quad \dot{z}_0^N = P_N v_0,$$

where $f_N = P_N f$. We can obtain the mild solution of (3.57) as

$$
\begin{aligned}
z^N(t) =& \cos\left(A^{\frac{\alpha}{2}}t\right) u_0^N + A^{-\frac{\alpha}{2}} \sin\left(A^{\frac{\alpha}{2}}t\right) v_0^N \\
& + \int_0^t A^{-\frac{\alpha}{2}} \sin\left(A^{\frac{\alpha}{2}}(t-s)\right) f_N\left(u^N(s)\right) ds, \\
\dot{z}^N(t) =& -A^{\frac{\alpha}{2}} \sin\left(A^{\frac{\alpha}{2}}t\right) u_0^N + \cos\left(A^{\frac{\alpha}{2}}t\right) v_0^N \\
& + \int_0^t \cos\left(A^{\frac{\alpha}{2}}(t-s)\right) f_N\left(u^N(s)\right) ds.
\end{aligned}
\tag{3.58}
$$

Then the spatial semi-discretization solution of (3.44) is given by

$$u^N(t) = z^N(t) + P_N \int_0^t A^{-\frac{\alpha}{2}} \sin\left(A^{\frac{\alpha}{2}}(t-s)\right) dB_H(s) \tag{3.59}$$

and

$$v^N(t) = \dot{z}^N(t) + P_N \int_0^t \cos\left(A^{\frac{\alpha}{2}}(t-s)\right) dB_H(s). \tag{3.60}$$

In fact, Corollary 3.2 shows the mild solution $z(t)$ of (3.52) has better regularity than the stochastic integral $\int_0^t A^{-\frac{\alpha}{2}} \sin\left(A^{\frac{\alpha}{2}}(t-s)\right)$ $dB_H(s)$ in space, so we can improve accuracy of the Galerkin approximate solution by postprocessing the stochastic integral of (3.59). Let $N_1 = [N^{\theta_3}]$ and $\theta_3 \geq 1$, with $[y]$ being the nearest integer to y. Then the spatial semi-discretization solution of (3.44) can be expressed as

$$u^N(t) = z^N(t) + P_{N_1} \int_0^t A^{-\frac{\alpha}{2}} \sin\left(A^{\frac{\alpha}{2}}(t-s)\right) dB_H(s) \tag{3.61}$$

and

$$v^N(t) = \dot{z}^N(t) + P_{N_1} \int_0^t \cos\left(A^{\frac{\alpha}{2}}(t-s)\right) dB_H(s). \tag{3.62}$$

Using (3.57) leads to

$$
\begin{aligned}
\dot{z}^N(t) =& -A^{\frac{\alpha}{2}} \sin\left(A^{\frac{\alpha}{2}}(t-s)\right) z^N(s) + \cos\left(A^{\frac{\alpha}{2}}(t-s)\right) \dot{z}^N(s) \\
& + \int_s^t \cos\left(A^{\frac{\alpha}{2}}(t-r)\right) f_N\left(u^N(r)\right) dr.
\end{aligned}
\tag{3.63}
$$

Then substituting (3.61) and (3.62) into (3.63) leads to

$$
\begin{aligned}
v^N(t) &= -A^{\frac{\alpha}{2}} \sin\left(A^{\frac{\alpha}{2}}(t-s)\right) u^N(s) + \cos\left(A^{\frac{\alpha}{2}}(t-s)\right) v^N(s) \\
&\quad + P_{N_1} \int_s^t \cos\left(A^{\frac{\alpha}{2}}(t-r)\right) \mathrm{d}B_H(r) \\
&\quad + \int_s^t \cos\left(A^{\frac{\alpha}{2}}(t-r)\right) f_N\left(u^N(r)\right) \mathrm{d}r,
\end{aligned} \tag{3.64}
$$

which will be used to prove Theorem 3.8.

Combining Theorem 3.6, (3.61), and (3.62), we now deduce the following regularity results of the spatial semi-discretization solution.

Corollary 3.3 *Suppose that Assumption 1 and 2 are satisfied;* $\|u_0\|_{L^p(D,\dot{U}^\gamma)} < \infty$, $\|v_0\|_{L^p(D,\dot{U}^{\gamma-\alpha})} < \infty$, $\varepsilon > 0$, $\gamma = \alpha + 2\rho - \frac{d+\varepsilon}{2}$ *and* $\gamma > 0$. *The approximate solutions* $u^N(t)$ *and* $v^N(t)$ *are expressed by (3.61) and (3.62), respectively. Then*

$$
\left\|u^N(t)\right\|_{L^p(D,\dot{U}^\gamma)} + \left\|v^N(t)\right\|_{L^p(D,\dot{U}^{\gamma-\alpha})} \lesssim \frac{t^H}{\varepsilon} + \|u_0\|_{L^p(D,\dot{U}^\gamma)}
$$

$$
+ \|v_0\|_{L^p(D,\dot{U}^{\gamma-\alpha})}
$$

and

(i) *for* $\gamma \leq \alpha$,

$$
\left\|u^N(t) - u^N(s)\right\|_{L^2(D,U)}
$$

$$
\lesssim (t-s)^{\frac{\gamma}{\alpha}} \left(\frac{t^H}{\varepsilon} + \|u_0\|_{L^2(D,\dot{U}^\gamma)} + \|v_0\|_{L^2(D,\dot{U}^{\gamma-\alpha})} \right).
$$

(ii) *for* $\gamma > \alpha$,

$$
\left\|u^N(t) - u^N(s)\right\|_{L^2(D,U)}
$$

$$
\lesssim (t-s) \left(t^H + \|u_0\|_{L^2(D,\dot{U}^\gamma)} + \|v_0\|_{L^2(D,\dot{U}^{\gamma-\alpha})} \right).
$$

The proof of this corollary is done in the same way as Theorem 3.6.

Next, the following lemma is given to analyze the error of the approximate solution $u^N(t)$ in (3.61).

Lemma 3.5 *If* $\mathrm{E}\left[\|A^{\frac{\nu}{2}}\xi\|^2\right] < \infty$, $\xi \in U$, *then*

$$
\mathrm{E}\left[\|(P_N - I)\xi\|^2\right] \lesssim \lambda_{N+1}^{-\nu} \mathrm{E}\left[\|A^{\frac{\nu}{2}}\xi\|^2\right].
$$

Corollary 3.2 shows $z(t) \in L^p(D, \dot{U}^{\gamma+\alpha})$. According to Lemma 3.5, we can infer that $\mathrm{E}\left[\|z(t) - P_N z(t)\|^2\right] \lesssim \lambda_{N+1}^{-\gamma-\alpha} \mathrm{E}\left[\|A^{\frac{\gamma+\alpha}{2}} z(t)\|^2\right]$. Therefore, we can obtain an order of $\frac{\gamma+\alpha}{d}$ for the spatial semi-discretization solution by adjusting N_1 in (3.61). Let $N_1 = \left[N^{\frac{\gamma+\alpha}{\gamma}}\right]$. Then we get the following result.

Theorem 3.7 *Let $X(t)$ and $Z^N(t)$ be the mild solutions of (3.44) and (3.52), respectively. Suppose that Assumptions 1 and 2 are satisfied. Let $\|u_0\|_{L^p(D,\dot{U}^{\gamma+\alpha})} < \infty$, $\|v_0\|_{L^p(D,\dot{U}^{\gamma})} < \infty$, $\varepsilon > 0$, $\gamma = \alpha + 2\rho - \frac{d+\varepsilon}{2}$, $\gamma > 0$; and (3.61) is the approximation of $u(t)$. If $N_1 = \left[N^{\frac{\gamma+\alpha}{\gamma}}\right]$, then we have*

$$\left\|u(t) - u^N(t)\right\|_{L^2(D,U)}$$
$$\lesssim N^{-\frac{\gamma+\alpha}{d}} \left(\frac{t^H}{\varepsilon} + \|u_0\|_{L^2(D,\dot{U}^{\gamma+\alpha})} + \|v_0\|_{L^2(D,\dot{U}^{\gamma})}\right).$$

Proof: In the first place, using the triangle inequality, Corollary 3.2, Lemma 3.5, and (3.61), we obtain

$$\left\|u(t) - u^N(t)\right\|_{L^2(D,U)}$$
$$\lesssim \left\|z(t) - z^N(t)\right\|_{L^2(D,U)}$$
$$+ \left\|\int_0^t \sum_{i=N_1+1}^{\infty} \lambda_i^{-\frac{\alpha}{2}} \sin\left(\lambda_i^{\frac{\alpha}{2}}(t_m - s)\right) \sigma_i \phi_i(x) \mathrm{d}\beta_H^i(s)\right\|_{L^2(D,U)}$$
$$\lesssim \|z(t) - P_N z(t)\|_{L^2(D,U)} + \left\|P_N z(t) - z^N(t)\right\|_{L^2(D,U)} + \frac{t^H}{\varepsilon} N_1^{-\frac{\gamma}{d}}$$
$$\lesssim \left\|P_N z(t) - z^N(t)\right\|_{L^2(D,U)}$$
$$+ N^{-\frac{\gamma+\alpha}{d}} \left(\frac{t^H}{\varepsilon} + \|u_0\|_{L^2(D,\dot{U}^{\gamma+\alpha})} + \|v_0\|_{L^2(D,\dot{U}^{\gamma})}\right).$$
$$(3.65)$$

Then we need to estimate the bound of $\left\|P_N z(t) - z^N(t)\right\|_{L^2(D,U)}$. The definition of projection operator P_N implies that $P_N A^{-\frac{\alpha}{2}} \sin\left(A^{\frac{\alpha}{2}} t\right) f = A^{-\frac{\alpha}{2}} \sin\left(A^{\frac{\alpha}{2}} t\right) P_N f$. Thus, first performing P_N on

(3.54) and then doing subtraction with respect to (3.58) lead to

$$\left\| P_N z(t) - z^N(t) \right\|_{L^2(D,U)}$$

$$= \left\| \int_0^t A^{-\frac{\alpha}{2}} \sin\left(A^{\frac{\alpha}{2}}(t_{m+1} - s)\right)\left(f_N\left(u(s)\right) - f_N\left(u^N(s)\right)\right) ds \right\|_{L^2(D,U)}$$

$$\lesssim \int_0^t \left\| u(s) - u^N(s) \right\|_{L^2(D,U)} ds.$$

Then the above estimates and the Grönwall inequality imlpy

$$\left\| u(t) - u^N(t) \right\|_{L^2(D,U)}$$

$$\lesssim N^{-\frac{\gamma+\alpha}{d}} \left(\frac{t^H}{\varepsilon} + \|u_0\|_{L^2(D,\dot{U}^{\gamma+\alpha})} + \|v_0\|_{L^2(D,\dot{U}^\gamma)} \right). \qquad (3.66)$$

If $N > 1$, choosing $\varepsilon = \frac{1}{\log(N)}$, we have

$$\left\| u(t) - u^N(t) \right\|_{L^2(D,U)}$$

$$\lesssim N^{-\frac{4\alpha+4\rho-d}{2d}} \left(t^H \log(N) + \|u_0\|_{L^2(D,\dot{U}^{\gamma+\alpha})} + \|v_0\|_{L^2(D,\dot{U}^\gamma)} \right).$$

$$\square$$

3.3.1.3 Fully discrete scheme

In this section, we concern the time discretization of (3.57). Meanwhile the error estimates of the fully discrete scheme are derived.

Let z_m^N and $\bar{z}_m^N$ denote respectively the approximation of $z^N(t_m)$ and $\dot{z}^N(t_m)$ with fixed time step size $\tau = \frac{T}{M}$ and $t_m = m\tau$ ($m = 0, 1, 2, \ldots, M$). Using the stochastic trigonometric method, we can get the fully discrete scheme of (3.44)

$$\begin{bmatrix} z_{m+1}^N \\ \bar{z}_{m+1}^N \end{bmatrix} = \begin{bmatrix} \cos\left(A^{\frac{\alpha}{2}}\tau\right) & A^{-\frac{\alpha}{2}}\sin\left(A^{\frac{\alpha}{2}}\tau\right) \\ -A^{\frac{\alpha}{2}}\sin\left(A^{\frac{\alpha}{2}}\tau\right) & \cos\left(A^{\frac{\alpha}{2}}\tau\right) \end{bmatrix} \begin{bmatrix} z_m^N \\ \bar{z}_m^N \end{bmatrix} \qquad (3.67)$$

$$+ \tau \begin{bmatrix} A^{-\frac{\alpha}{2}}\sin\left(A^{\frac{\alpha}{2}}\tau\right) f_N\left(u_m^N\right) \\ \cos\left(A^{\frac{\alpha}{2}}\tau\right) f_N\left(u_m^N\right) \end{bmatrix},$$

where

$$u_m^N = z_m^N + P_{N_1} \int_0^{t_m} A^{-\frac{\alpha}{2}} \sin\left(A^{\frac{\alpha}{2}}(t_m - s)\right) dB_H(s).$$

By using recursion form of (3.67), we get

$$
\begin{bmatrix} z_{m+1}^N \\ \tilde{z}_{m+1}^N \end{bmatrix} = \begin{bmatrix} \cos\left(A^{\frac{\alpha}{2}} t_{m+1}\right) & A^{-\frac{\alpha}{2}} \sin\left(A^{\frac{\alpha}{2}} t_{m+1}\right) \\ -A^{\frac{\alpha}{2}} \sin\left(A^{\frac{\alpha}{2}} t_{m+1}\right) & \cos\left(A^{\frac{\alpha}{2}} t_{m+1}\right) \end{bmatrix} \begin{bmatrix} u_0^N \\ v_0^N \end{bmatrix}
$$

$$
+ \tau \sum_{j=0}^{m} \begin{bmatrix} A^{-\frac{\alpha}{2}} \sin\left(A^{\frac{\alpha}{2}}(t_{m+1} - t_j)\right) f_N\left(u_j^N\right) \\ \cos\left(A^{\frac{\alpha}{2}}(t_{m+1} - t_j)\right) f_N\left(u_j^N\right) \end{bmatrix}.
$$

$$(3.68)$$

Then we get the approximations of $u(t)$ and $v(t)$, that is,

$$
u_{m+1}^N = z_{m+1}^N + P_{N_1} \int_0^{t_{m+1}} A^{-\frac{\alpha}{2}} \sin\left(A^{\frac{\alpha}{2}}(t_{m+1} - s)\right) \mathrm{d}B_H(s) \quad (3.69)
$$

and

$$
v_{m+1}^N = \tilde{z}_{m+1}^N + P_{N_1} \int_0^{t_{m+1}} \cos\left(A^{\frac{\alpha}{2}}(t_{m+1} - s)\right) \mathrm{d}B_H(s). \quad (3.70)
$$

As $\frac{1}{2} < H < 1$, although $\int_0^{t_{m+1}} \sin\left(A^{\frac{\alpha}{2}}(t_{m+1} - s)\right) \mathrm{d}B_H(s)$ is a Gaussian process, it is difficult to accurately simulate this process; thus we give the approximation of stochastic integral, that is

$$
P_{N_1} \sum_{j=0}^{m} \int_{t_j}^{t_{j+1}} A^{-\frac{\alpha}{2}} \sin\left(A^{\frac{\alpha}{2}}(t_{m+1} - t_j)\right) \mathrm{d}B_H(s). \quad (3.71)
$$

Using Lemma 3.2, we obtain the error estimate

$$
\mathrm{E}\left[\left\|P_{N_1} \sum_{j=0}^{m} \int_{t_j}^{t_{j+1}} A^{-\frac{\alpha}{2}} \left(\sin\left(A^{\frac{\alpha}{2}}(t_{m+1} - s)\right) \right.\right.\right.
$$

$$
\left.\left.\left. - \sin\left(A^{\frac{\alpha}{2}}(t_{m+1} - t_j)\right)\right) \mathrm{d}B_H(s)\right\|^2\right]
$$

$$
\lesssim t_{m+1}^{2H} \tau^{\min\left\{\frac{2\gamma}{\alpha}, 2\right\}} \sum_{i=0}^{N_1} \lambda_i^{\min\{\gamma, \alpha\} - \alpha - 2\rho},
$$

which implies that this approximation (3.71) does not change the temporal convergence rate of scheme (3.67). The proof of this estimate is done in the same way as (3.76). For $H = \frac{1}{2}$, the simulation of stochastic integral is easily implementable without approximation.

We now investigate the error estimates of the fully discrete scheme (3.67). The triangle inequality implies that

$$\left\|u(t_m) - u_m^N\right\|_{L^2(D,U)}$$
$$\lesssim \left\|u(t_m) - u^N(t_m)\right\|_{L^2(D,U)} + \left\|u^N(t_m) - u_m^N\right\|_{L^2(D,U)}.$$

Thus, we need to give the bound estimate of $\left\|u^N(t_m) - u_m^N\right\|_{L^2(D,U)}$. This bound estimate can be obtained by using the time Hölder regularity of $u^N(t)$, (3.61), and (3.69). Therefore combining the error estimate of the approximation (3.71), Corollary 3.3, and Theorem 3.7 leads to the following results.

Proposition 3.1 *Let $u(t_{m+1})$ and u_{m+1}^N be expressed by (3.47) and (3.69), respectively. Suppose that Assumptions 1 and 2 are satisfied. Suppose Corollary 3.3 and Theorem 3.7 hold. Let $N_1 = \left[N^{\frac{\gamma+\alpha}{\gamma}}\right]$. Then*

$$\left\|u(t_{m+1}) - u_{m+1}^N\right\|_{L^2(D,U)}$$
$$\lesssim \tau \left(T^H + \|u_0\|_{L^2(D,\dot{U}^{\gamma+\alpha})} + \|v_0\|_{L^2(D,\dot{U}^\gamma)}\right)$$
$$+ N^{-\frac{\gamma+\alpha}{d}} \left(\frac{T^H}{\varepsilon} + \|u_0\|_{L^2(D,\dot{U}^{\gamma+\alpha})} + \|v_0\|_{L^2(D,\dot{U}^\gamma)}\right), \gamma > \alpha$$

and

$$\left\|u(t_{m+1}) - u_{m+1}^N\right\|_{L^2(D,U)}$$
$$\lesssim \tau^{\frac{\gamma}{\alpha}} \left(\frac{T^H}{\varepsilon} + \|u_0\|_{L^2(D,\dot{U}^{\gamma+\alpha})} + \|v_0\|_{L^2(D,\dot{U}^\gamma)}\right)$$
$$+ N^{-\frac{\gamma+\alpha}{d}} \left(\frac{T^H}{\varepsilon} + \|u_0\|_{L^2(D,\dot{U}^{\gamma+\alpha})} + \|v_0\|_{L^2(D,\dot{U}^\gamma)}\right), \gamma \le \alpha.$$

As $\gamma > \alpha$, the derivative of $u(t)$ is time Hölder continuous in the sense of mean-squared L^p-norm, which means that the scheme (3.67) is not optimal to discretize (3.43) in time. Thus we can design a higher order scheme for the time discretization, as $v(t)$ belongs to $L^p(D, U)$. By modifying the scheme (3.67), we can get a better convergence rate than one of (3.67) in time. The modified scheme is as follows:

$$
\begin{bmatrix} z_1^N \\ \bar{z}_1^N \end{bmatrix} = \begin{bmatrix} \cos\left(A^{\frac{\alpha}{2}}\tau\right) & A^{-\frac{\alpha}{2}}\sin\left(A^{\frac{\alpha}{2}}\tau\right) \\ -A^{\frac{\alpha}{2}}\sin\left(A^{\frac{\alpha}{2}}\tau\right) & \cos\left(A^{\frac{\alpha}{2}}\tau\right) \end{bmatrix} \begin{bmatrix} z_0^N \\ \bar{z}_0^N \end{bmatrix} \tag{3.72}
$$
$$
+ \begin{bmatrix} A^{-\alpha}\left(1 - \cos\left(A^{\frac{\alpha}{2}}\tau\right)\right) f_N\left(u_0^N\right) \\ A^{-\frac{\alpha}{2}}\sin\left(A^{\frac{\alpha}{2}}\tau\right) f_N\left(u_0^N\right) \end{bmatrix}
$$

and for $m \geq 1$,

$$
\begin{bmatrix} z_{m+1}^N \\ \bar{z}_{m+1}^N \end{bmatrix} = \begin{bmatrix} \cos\left(A^{\frac{\alpha}{2}}\tau\right) & A^{-\frac{\alpha}{2}}\sin\left(A^{\frac{\alpha}{2}}\tau\right) \\ -A^{\frac{\alpha}{2}}\sin\left(A^{\frac{\alpha}{2}}\tau\right) & \cos\left(A^{\frac{\alpha}{2}}\tau\right) \end{bmatrix} \begin{bmatrix} z_m^N \\ \bar{z}_m^N \end{bmatrix}
$$

$$
\tag{3.73}
$$

$$
+ \begin{bmatrix} A^{-\alpha}\left(1 - \cos\left(A^{\frac{\alpha}{2}}\tau\right)\right) f_N\left(u_m^N\right) \\ A^{-\frac{\alpha}{2}}\sin\left(A^{\frac{\alpha}{2}}\tau\right) f_N\left(u_m^N\right) \end{bmatrix}
$$

$$
+ \begin{bmatrix} \frac{\tau A^{-\alpha} - A^{-\frac{3\alpha}{2}}\sin\left(A^{\frac{\alpha}{2}}\tau\right)}{\tau}\left(f_N\left(u_m^N\right) - f_N\left(u_{m-1}^N\right)\right) \\ \frac{A^{-\alpha} - A^{-\alpha}\cos\left(A^{\frac{\alpha}{2}}\tau\right)}{\tau}\left(f_N\left(u_m^N\right) - f_N\left(u_{m-1}^N\right)\right) \end{bmatrix}.
$$

Then a classic application of recursion gives

$$
u_{m+1}^N = \cos\left(A^{\frac{\alpha}{2}}t_{m+1}\right) u_0^N + A^{-\frac{\alpha}{2}}\sin\left(A^{\frac{\alpha}{2}}t_{m+1}\right) v_0^N \tag{3.74}
$$

$$
+ \sum_{j=0}^{m} A^{-\alpha}\left(\cos\left(A^{\frac{\alpha}{2}}(t_{m+1} - t_{j+1})\right)\right.
$$

$$
\left. - \cos\left(A^{\frac{\alpha}{2}}(t_{m+1} - t_j)\right)\right) f_N\left(u_j^N\right)
$$

$$
+ \sum_{j=1}^{m} A^{-\alpha}\cos\left(A^{\frac{\alpha}{2}}(t_{m+1} - t_{j+1})\right)\left(f_N\left(u_j^N\right) - f_N\left(u_{j-1}^N\right)\right)
$$

$$
- \sum_{j=1}^{m} \frac{A^{-\alpha}\int_{t_j}^{t_{j+1}}\cos\left(A^{\frac{\alpha}{2}}(t_{m+1} - s)\right)ds}{\tau}
$$

$$
\times \left(f_N\left(u_j^N\right) - f_N\left(u_{j-1}^N\right)\right)
$$

$$
+ P_{N_1}\int_0^{t_{m+1}} A^{-\frac{\alpha}{2}}\sin\left(A^{\frac{\alpha}{2}}(t_{m+1} - s)\right)dB_H(s)
$$

and

$$v_{m+1}^N = - A^{\frac{\alpha}{2}} \sin\left(A^{\frac{\alpha}{2}} t_{m+1}\right) u_0^N + \cos\left(A^{\frac{\alpha}{2}} t_{m+1}\right) v_0^N \qquad (3.75)$$

$$- \sum_{j=0}^{m} A^{-\frac{\alpha}{2}} \left(\sin\left(A^{\frac{\alpha}{2}}(t_{m+1} - t_{j+1})\right)\right.$$

$$\left. - \sin\left(A^{\frac{\alpha}{2}}(t_{m+1} - t_j)\right)\right) f_N\left(u_j^N\right)$$

$$- \sum_{j=1}^{m} A^{-\frac{\alpha}{2}} \sin\left(A^{\frac{\alpha}{2}}(t_{m+1} - t_{j+1})\right) \left(f_N\left(u_j^N\right) - f_N\left(u_{j-1}^N\right)\right)$$

$$+ \sum_{j=1}^{m} \frac{A^{-\frac{\alpha}{2}} \int_{t_j}^{t_{j+1}} \sin\left(A^{\frac{\alpha}{2}}(t_{m+1} - s)\right) ds}{\tau}$$

$$\times \left(f_N\left(u_j^N\right) - f_N\left(u_{j-1}^N\right)\right)$$

$$+ P_{N_1} \int_0^{t_{m+1}} \cos\left(A^{\frac{\alpha}{2}}(t_{m+1} - s)\right) dB_H(s).$$

For $\frac{1}{2} < H < 1$ and $\gamma > \alpha$, if the time steps of (3.71) and (3.73) are the same, the desired convergence rate cannot be got. Thus, a more precise approximation of stochastic integral than (3.71) is expected. The inequality $\cos(t_{m+1} - r) - \cos(t_{m+1} - t_j) \lesssim |r - t_j|^\theta$ ($0 \le \theta \le 1$) and the error equation $\int_{t_j}^{t_{j+1}} \int_{t_j}^{s} \cos\left(A^{\frac{\alpha}{2}}(t_{m+1} - r)\right) dr dB_H(s)$ of (3.71) imply that we can improve the accuracy of approximation for stochastic integral by the following scheme, that is,

$$P_{N_1} \sum_{j=0}^{m} \int_{t_j}^{t_{j+1}} \left(A^{-\frac{\alpha}{2}} \sin\left(A^{\frac{\alpha}{2}}(t_{m+1} - t_j)\right)\right.$$

$$\left. - (s - t_j) \cos\left(A^{\frac{\alpha}{2}}(t_{m+1} - t_j)\right)\right) dB_H(s), \qquad (3.76)$$

which ensures the implementation of scheme (3.73) without loss of convergence rate; and Equation (3.76) is easy to simulate by using its explicit variance. Using Lemma 3.2, the following error estimate of the approximation for stochastic integral is obtained

$$
\mathrm{E}\left[\left\|P_{N_1}\sum_{j=0}^{m}\int_{t_j}^{t_{j+1}}\int_{t_j}^{s}\left(\cos\left(A^{\frac{\alpha}{2}}(t_{m+1}-t_j)\right)\right.\right.\right.
$$
$$
\left.\left.\left.-\cos\left(A^{\frac{\alpha}{2}}(t_{m+1}-r_1)\right)\right)\,\mathrm{d}r_1\mathrm{d}B_H(s)\right\|^2\right]
$$
$$
\lesssim\tau^{\min\left\{\frac{2\gamma}{\alpha},4\right\}}\sum_{i=0}^{N_1}\lambda_i^{\min\{\gamma-\alpha,\alpha\}-2\rho}\sum_{j=0}^{m}\int_{t_j}^{t_{j+1}}\sum_{k=0}^{m}\int_{t_k}^{t_{k+1}}|s-t|^{2H-2}\mathrm{d}s\mathrm{d}t
$$
$$
=\tau^{\min\left\{\frac{2\gamma}{\alpha},4\right\}}\sum_{i=0}^{N_1}\lambda_i^{\min\{\gamma-\alpha,\alpha\}-2\rho}\int_{0}^{t_{m+1}}\int_{0}^{t_{m+1}}|s-t|^{2H-2}\mathrm{d}s\mathrm{d}t
$$
$$
\lesssim t_{m+1}^{2H}\tau^{\min\left\{\frac{2\gamma}{\alpha},4\right\}}\sum_{i=0}^{N_1}\lambda_i^{\min\{\gamma-\alpha,\alpha\}-2\rho},
$$

$$(3.77)$$

the second inequality of which uses the fact

$$
\cos\left(\lambda_i^{\frac{\alpha}{2}}(t_{m+1}-s)\right)-\cos\left(\lambda_i^{\frac{\alpha}{2}}(t_{m+1}-t)\right)
$$
$$
\lesssim\lambda_i^{\min\left\{\frac{\gamma-\alpha}{2},\frac{\alpha}{2}\right\}}|s-t|^{\min\left\{\frac{\gamma-\alpha}{\alpha},1\right\}}.
$$

We end this section by showing the error estimates of the fully discrete scheme (3.73) in $L^2(D,U)$ norm.

Theorem 3.8 *Let $u(t_{m+1})$ and u_{m+1}^N be expressed by (3.47) and (3.74), respectively. Suppose that $f(u)\in C^1(\mathbb{R})$ and $f'(u)$ satisfies the Lipschitz condition, and the conditions of Corollary 3.3 and Theorem 3.7 are satisfied. Take $N>1$ and $0<\tau<1$. If $\gamma>\alpha$ and $N_1=\left[N^{\frac{\gamma+\alpha}{\gamma}}\right]$, then*
(i) *for $\alpha<\gamma\leq2\alpha$,*

$$
\left\|u(t_{m+1})-u_{m+1}^N\right\|_{L^2(D,U)}
$$
$$
\lesssim\tau^{\frac{\alpha+2\rho-\frac{d}{2}}{\alpha}}\left(T^H|\log(\tau)|+\|u_0\|_{L^2(D,\dot{U}^{\gamma+\alpha})}+\|v_0\|_{L^2(D,\dot{U}^\gamma)}\right)
$$
$$
+N^{-\frac{2\rho+2\alpha-\frac{d}{2}}{d}}\left(T^H\log(N)+\|u_0\|_{L^2(D,\dot{U}^{\gamma+\alpha})}+\|v_0\|_{L^2(D,\dot{U}^\gamma)}\right);
$$

(ii) *for* $\gamma > 2\alpha$,

$$\left\| u(t_{m+1}) - u_{m+1}^N \right\|_{L^2(D,U)}$$

$$\lesssim \tau^2 \left(T^H + \|u_0\|_{L^2(D,\dot{U}^{\gamma+\alpha})} + \|v_0\|_{L^2(D,\dot{U}^\gamma)} \right)$$

$$+ N^{-\frac{2\rho+2\alpha-\frac{d}{2}}{d}} \left(T^H \log(N) + \|u_0\|_{L^2(D,\dot{U}^{\gamma+\alpha})} + \|v_0\|_{L^2(D,\dot{U}^\gamma)} \right).$$

Proposition 3.1 and Theorem 3.8 show that one can choose the appropriate technique to solve (3.42), i.e., when $\alpha \le \gamma$, use (3.67) to discretize (3.42), and if $\alpha > \gamma$, the scheme (3.73) can be chosen to obtain the approximation of $u(t)$. In fact, when $u_0 \in L^2(D, \dot{U}^\gamma)$ and $v_0 \in L^2(D, \dot{U}^{\gamma-\alpha})$, the temporal rates of convergence still hold in Proposition 3.1 and Theorem 3.8.

3.3.1.4 *Numerical experiments*

In this section, we present numerical examples to verify the theoretical results and the effect of the parameters α and ρ on the convergence. All numerical errors are given in the sense of mean-squared L^2-norm.

We solve (3.42) in the two-dimensional domain $D = (0,1) \times (0,1)$ by the proposed scheme (3.73) with $x = (x_1, x_2)$, the smooth initial data $v_0 = \sin(4\pi x_1)\sin(4\pi x_2)$, and $u_0 = \frac{\sin(\pi x_1)\sin(\pi x_2)}{2}$. In $D = (0,1) \times (0,1)$, the Dirichlet eigenpairs of $-\Delta$ are $\lambda_{i,j} = \pi^2(i^2 + j^2)$ and $\phi_{i,j} = 2\sin(i\pi x_1)\sin(j\pi x_2)$ with $i, j = 1, 2, \ldots, N_0$ ($N = N_0^2$). Unless otherwise specified, we choose $f(u(t)) = u(t)$. We take 1000 as the number of the simulation trajectories.

Table 3.6 Spatial convergence rates with $T = 0.3$, $M = 900$, $H = 0.5$, and $\rho = 1$

	N	$\alpha = 0.4$	Rate	$\alpha = 0.6$	Rate	$\alpha = 0.8$	Rate
$N_1 = N$	256	2.610e-04		1.064e-04		4.810e-05	
	576	1.608e-04	0.597	5.910e-05	0.725	2.365e-05	0.875
	1296	9.545e-05	0.643	3.220e-05	0.749	1.173e-05	0.865
$N_1 = \left\lceil N^{\frac{\gamma+\alpha}{\gamma}} \right\rceil$	256	1.051e-04		2.367e-05		5.912e-06	
	576	4.787e-05	0.970	9.867e-06	1.079	2.073e-06	1.292
	1296	2.175e-05	0.973	4.055e-06	1.097	7.343e-07	1.280

The spatial convergence rates of the scheme (3.73) are tested with the end time $T = 0.3$ and $M = 900$, which ensures the spatial error is the dominant one. In Table 3.6, one can see that the spatial convergence rates tend to $\frac{2\rho+\alpha-1}{2}$ if $N_1 = N$, and the convergence rates are approximately equal to $\frac{2\rho+2\alpha-1}{2}$ after postprocessing the stochastic integral $\left(N_1 = \left[N^{\frac{\gamma+\alpha}{\gamma}}\right]\right)$. And the convergence rates of the spectral Galerkin method are improved, as α increases. The numerical results verify the theoretical ones.

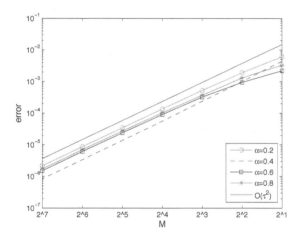

Figure 3.1 Temporal error convergence of the modified stochastic trigonometric method for the space-time white noise ($H = 0.5$).

Next, we observe the behavior of the temporal convergence. We solve the problem (3.42) by using the scheme (3.73) with $f(u(t)) = \sin(u(t))$, $\rho = 2.5$, $T = 0.6$, and $N = 400$ in Figures 3.1 and 3.2. The sufficiently big ρ and N guarantee that the dominant errors arise from the temporal approximation. As $H = \frac{1}{2}$, the simulation of the stochastic integral $\int_0^T \sin(\lambda_{i,j}(T - t))\,\mathrm{d}\beta(t)$ is easily implementable by using explicit variance of the stochastic integral, which is $\frac{T}{2} - \frac{\sin(2\lambda_{i,j}T)}{4\lambda_{i,j}}$. For $H \in \left(\frac{1}{2}, 1\right)$, one can obtain the approximation of the stochastic integral by using scheme (3.76). Figures 3.1 and 3.2 show that the temporal convergence rates have an order of 2 by using the proposed scheme, as $\gamma > 2\alpha$, and the convergence rates are independent of H.

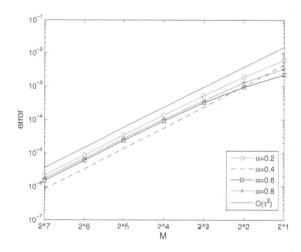

Figure 3.2 Temporal error convergence of the modified stochastic trigonometric method for the space-time fractional Gaussian noise $(H = 0.75)$.

Table 3.7 Time convergence rates with $N = 10^6$, $T - 0.6$, $H - 0.5$, and $\rho = 0.68$

M	$\alpha = 0.5$	Rate	$\alpha = 0.7$	Rate	$\alpha = 0.9$	Rate
4	1.400e-03		2.414e-03		2.980e-03	
8	4.158e-04	1.751	8.575e-04	1.493	1.134e-03	1.394
16	1.220e-04	1.769	3.077e-04	1.479	4.341e-04	1.385

As $\alpha < \gamma \leq 2\alpha$, the convergence rates of the proposed scheme are close to $\frac{\alpha+2\rho-1}{\alpha}$ in time. For $H = \frac{1}{2}$, from Table 3.7, one can see that the temporal convergence rates reduce with the increase of α, for fixing ρ. As $\alpha = 0.5$, 0.7, and 0.9, the theoretical convergence rates are approximately 1.720, 1.514, and 1.400, respectively. For $H = 0.6$, Table 3.8 demonstrates that the time convergence rates increase with the increase of ρ, for fixing α. The temporal theoretical convergence rates are approximately 1.556, 1.667, and 1.778, for $\rho = 0.75$, 0.80, and 0.85, respectively. Tables 3.7 and 3.8 show that the numerical results confirm the error estimates in Theorem 3.8.

Table 3.8 Time convergence rates with $N = 10^6$, $T = 0.6$, $H = 0.6$, and $\alpha = 0.9$

M	$\rho = 0.75$	Rate	$\rho = 0.8$	Rate	$\rho = 0.85$	Rate
4	4.568e-03		3.543e-03		2.907e-03	
8	1.559e-03	1.551	1.156e-03	1.616	8.956e-04	1.699
16	5.330e-04	1.548	3.733e-04	1.630	2.704e-04	1.728

3.3.2 Galerkin Finite Element Approximation of Stochastic Fractional Wave Equations

The authors of [115] provide the finite element approximation for the following initial boundary value problem with $1 < \alpha < 2$ and $\frac{1}{2} < \beta \le 1$

$$\begin{cases} \partial_t^\alpha u(x,t) = -(-\Delta)^\beta u(x,t) + \frac{\partial^2 W(x,t)}{\partial t \partial x} & \text{in } D \times (0,T], \\ u(x,0) = u_0(x), \ \dot{u}(x,0) = v_0(x) & \text{in } D, \\ u(x,t) = 0, \quad \partial D \times (0,T], \end{cases} \tag{3.78}$$

where $D = (0,1)$, ∂_t^α denotes the left-sided Caputo fractional derivative of order α with respect to t, and $W(x,t)$ represents an infinite dimensional Bm. Here the noise we consider is external noise. Suppose that $\frac{\partial^2 W(x,t)}{\partial t \partial x}$ is an infinite dimensional noise defined on $\left(\Omega, \mathcal{F}, \{\mathcal{F}_t\}_{t \in [0,T]}, P \right)$ such that

$$\frac{\partial^2 W(x,t)}{\partial t \partial x} = \sum_{i=1}^{\infty} \sigma_i(t) \dot{\xi}_i(t) \phi_i(x),$$

where $\sigma_i(t)$ is a continuous function, rapidly decaying with the increase of i to ensure the convergence of the series, and $\dot{\xi}_i(t) = \frac{d\xi_i(t)}{dt}$ is the white noise.

3.3.2.1 Solution representation

Now we define a partition of $[0,T]$ by intervals $[t_k, t_{k+1}]$ for $k = 0, 1, 2, \ldots, M$, where $t_k = k\Delta t$, $\Delta t = \frac{T}{M}$. A sequence of noise which approximates the space-time white noise is defined as [52]

$$\frac{\partial^2 W_m(x,t)}{\partial t \partial x} = \sum_{i=1}^{\infty} \sigma_i^m(t) \phi_i(x) \left(\sum_{k=0}^{M} \frac{1}{\sqrt{\Delta t}} \xi_{ik} \chi_k(t) \right),$$

where $\chi_k(t)$ is the characteristic function for the kth time subinterval,

$$\xi_{ik} = \frac{1}{\sqrt{\Delta t}} \int_k^{k+1} \mathrm{d}\xi_i(t) = \frac{1}{\sqrt{\Delta t}} (\xi_i(t_{k+1}) - \xi_i(t_k)) \sim N(0,1),$$

and $\sigma_i^m(t)$ is the approximation of $\sigma_i(t)$ in the space direction. More precisely, by replacing $\sigma_i(t)$ by $\sigma_i^m(t)$, we get the noise approximation in space, and by replacing $\dot{\xi}_i(t)$ by $\sum_{k=0}^M \frac{1}{\sqrt{\Delta t}} \xi_{ik} \chi_k(t)$, we get the noise approximation in time. Then $\frac{\partial^2 W_m(x,t)}{\partial t \partial x}$ is substituted for $\frac{\partial^2 W(x,t)}{\partial t \partial x}$ in Eqs. (3.78) to get the following equation:

$$\begin{cases} \partial_t^\alpha u_m(x,t) = -(-\Delta)^\beta u_m(x,t) + \frac{\partial^2 W_m(x,t)}{\partial t \partial x} & \text{in } D \times (0,T], \\ u_m(x,0) = u_0(x), \ \dot{u}_m(x,0) = v_0(x) & \text{in } D, \\ u_m(x,t) = 0, \quad \partial D \times (0,T], \end{cases}$$

$$(3.79)$$

The following lemma shows a representation of the mild solution to problem (3.78) using the Dirichlet eigenpairs $\{(\lambda_i, \phi_i(x))\}$.

Lemma 3.6 *The unique solution u to problem (3.78) with $1 < \alpha < 2$ and $\frac{1}{2} < \beta \le 1$ is given by*

$$u(x,t) = \int_0^1 \mathcal{T}_{\alpha,\beta}(x,y,t) u_0(y) \mathrm{d}y + \int_0^1 \mathcal{R}_{\alpha,\beta}(x,y,t) v_0(y) \mathrm{d}y$$
$$+ \int_0^1 \int_0^t \mathcal{S}_{\alpha,\beta}(x,y,t-s) \mathrm{d}W(y,s).$$

$$(3.80)$$

Here

$$\mathcal{T}_{\alpha,\beta}(x,y,t) = \sum_{i=1}^\infty E_{\alpha,1}(-\lambda_i^\beta t^\alpha) \phi_i(x) \phi_i(y) \qquad (3.81)$$

is the fundamental solution of

$$\partial_t^\alpha v(x,t) = -(-\Delta)^\beta v(x,t),$$
$$v(0,t) = v(1,t) = 0, \quad v(x,0) = \phi(x), \quad \partial_t v(x,0) = 0,$$

so that $v(x,t) = \int_0^1 \mathcal{T}_{\alpha,\beta}(x,y,t)\phi(y)\mathrm{d}y$, and

$$\mathcal{R}_{\alpha,\beta}(x,y,t) = \sum_{i=1}^{\infty} tE_{\alpha,2}(-\lambda_i^\beta t^\alpha)\phi_i(x)\phi_i(y) \qquad (3.82)$$

is the fundamental solution of

$$\partial_t^\alpha v(x,t) = -(-\Delta)^\beta v(x,t),$$

$$v(0,t) = v(1,t) = 0, \quad v(x,0) = 0, \quad \partial_t v(x,0) = \psi(x),$$

so that $v(x,t) = \int_0^1 \mathcal{R}_{\alpha,\beta}(x,y,t)\psi(y)\mathrm{d}y$. For (3.78) but with the initial data $v(x,0) = \partial_t v(x,0) \equiv 0$, we shall use the operator defined by

$$\mathcal{S}_{\alpha,\beta}(x,y,t) = \sum_{i=1}^{\infty} t^{\alpha-1}E_{\alpha,\alpha}(-\lambda_i^\beta t^\alpha)\phi_i(x)\phi_i(y) \qquad (3.83)$$

and $v(x,t) = \int_0^1 \int_0^t \mathcal{S}_{\alpha,\beta}(x,y,t-s)\mathrm{d}W(y,s)$.

Proof: Let

$$u(x,t) = \sum_{i=1}^{\infty} (u(t),\phi_i)\phi_i(x) = \sum_{i=1}^{\infty} u_i(t)\phi_i(x) \qquad (3.84)$$

be the solution of (3.78). Substituting (3.84) into (3.78), we get that for $1 < \alpha < 2$,

$$\partial_t^\alpha u_i(t) + \lambda_i^\beta u_i(t) = \sigma_i(t)\dot{\xi}_i(t), \quad u_i(0) = u_{0i}, \quad \partial_t u_i(0) = v_{0i}. \quad (3.85)$$

By Theorem 5.15 in [96], we have

$$u_i(t) = u_{0i}E_{\alpha,1}(-\lambda_i^\beta t^\alpha) + v_{0i}tE_{\alpha,2}(-\lambda_i^\beta t^\alpha)$$

$$+ \int_0^t (t-s)^{\alpha-1}E_{\alpha,\alpha}(-\lambda_i^\beta(t-s)^\alpha)\sigma_i(s)\dot{\xi}_i(s)\mathrm{d}s.$$

$$(3.86)$$

Then, it follows from (3.84) and (3.86) that

$$u(x,t) = \sum_{i=1}^{\infty} u_{0i} E_{\alpha,1}(-\lambda_i^{\beta} t^{\alpha}) \phi_i(x) + \sum_{i=1}^{\infty} v_{0k} t E_{\alpha,2}(-\lambda_i^{\beta} t^{\alpha}) \phi_i(x)$$

$$+ \sum_{i=1}^{\infty} \int_0^t (t-s)^{\alpha-1} E_{\alpha,\alpha}(-\lambda_i^{\beta}(t-s)^{\alpha}) \sigma_i(s) \dot{\xi}_k(s) ds \phi_i(x)$$

$$= \int_0^1 \sum_{i=1}^{\infty} E_{\alpha,1}(-\lambda_i^{\beta} t^{\alpha}) \phi_i(x) \phi_i(y) u_0(y) dy$$

$$+ \int_0^1 \sum_{i=1}^{\infty} t E_{\alpha,2}(-\lambda_i^{\beta} t^{\alpha}) \phi_i(x) \phi_i(y) v_0(y) dy$$

$$+ \int_0^t \int_0^1 (t-s)^{\alpha-1} \sum_{i=1}^{\infty} E_{\alpha,\alpha}(-\lambda_i^{\beta}(t-s)^{\alpha}) \phi_i(x) \phi_i(y) dW(s,y)$$

$$= \int_0^1 \mathcal{T}_{\alpha,\beta}(t,x,y) u_0(y) dy + \int_0^1 \mathcal{R}_{\alpha,\beta}(t,x,y) v_0(y) dy$$

$$+ \int_0^t \int_0^1 \mathcal{S}_{\alpha,\beta}(t-s,x,y) dW(s,y),$$

where we have used the fact that $E_{\alpha,\beta}(-\lambda_i^{\beta} t^{\alpha}) \sim [\lambda_i^{\beta} t^{\alpha} \Gamma(\beta-\alpha)]^{-1}$ when $i \to +\infty$, which ascertain the convergence of the above series.

Conversely, assume u satisfies (3.80). Then for $t - 0$,

$$u(0,x) = \int_0^1 \mathcal{T}_{\alpha,\beta}(0,x,y) u_0(y) dy = \int_0^1 \sum_{i=1}^{\infty} E_{\alpha,1}(0) \phi_i(x) \phi_i(y) u_0(y) dy$$

$$= \int_0^1 \sum_{i=1}^{\infty} \phi_i(x) \phi_i(y) u_0(y) dy = u_0(x),$$

and by Lemma 3.4 and Equation (1.10.7) in [96], we have

$$\partial_t u(0,x) = \partial_t \Big[\int_0^1 \mathcal{T}_{\alpha,\beta}(t,x,y) u_0(y) dy + \int_0^1 \mathcal{R}_{\alpha,\beta}(t,x,y) v_0(y) dy$$

$$+ \int_0^t \int_0^1 \mathcal{S}_{\alpha,\beta}(t-s,x,y) dW(s,y) \Big]_{t=0}$$

$$= v_0(x).$$

Further, to prove (3.80) satisfying (3.78) is equivalent to prove (3.80) satisfying

$$I_t^1 [\partial_t^\alpha u(x,t)] + I_t^1 [(-\Delta)^\beta u(x,t)] = I_t^1 \left[\frac{\partial^2 W(t,x)}{\partial t \partial x} \right]. \qquad (3.87)$$

Notice that

$$I_t^1 [\partial_t^\alpha u(x,t)] = \partial_t^{\alpha-1} u(x,t) - \frac{t^{2-\alpha}}{\Gamma(3-\alpha)} v_0(x).$$

Then applying $\partial_t^{\alpha-1}$ on both sides of (3.80), and using Lemma 3.4 and the Equation (1.100) in [150], we get

$$\partial_t^{\alpha-1} u(x,t) = \partial_t^{\alpha-1} \left[\int_0^1 \mathcal{T}_{\alpha,\beta}(t,x,y) u_0(y) dy \right.$$

$$+ \int_0^1 \mathcal{R}_{\alpha,\beta}(t,x,y) v_0(y) dy$$

$$\left. + \int_0^t \int_0^1 \mathcal{S}_{\alpha,\beta}(t-s,x,y) dW(s,y) \right]$$

$$= \int_0^1 \sum_{i=1}^\infty (-\lambda_i^\beta) t E_{\alpha,2}(-\lambda_i^\beta t^\alpha) \phi_i(x) \phi_i(y) u_0(y) dy$$

$$+ \int_0^1 \sum_{i=1}^\infty t^{2-\alpha} E_{\alpha,3-\alpha}(-\lambda_i^\beta t^\alpha) \phi_i(x) \phi_i(y) v_0(y) dy$$

$$+ \int_0^t \int_0^1 \sum_{i=1}^\infty E_{\alpha,1}(-\lambda_i^\beta (t-s)^\alpha) \phi_i(x) \phi_i(y) dW(s,y).$$

$$(3.88)$$

Applying $I_t^1(-\Delta)^\beta$ on both sides of (3.80), and using the Equation (1.100) in [150], we obtain

$$I_t^1[(-\Delta)^\beta u(x,t)]$$

$$= \int_0^t \int_0^1 \sum_{i=1}^\infty E_{\alpha,1}(-\lambda_i^\beta s^\alpha)\lambda_i^\beta \phi_i(x)\phi_i(y)u_0(y)dyds$$

$$+ \int_0^t \int_0^1 \sum_{i=1}^\infty sE_{\alpha,2}(-\lambda_i^\beta s^\alpha)\lambda_i^\beta \phi_i(x)\phi_i(y)v_0(y)dyds$$

$$+ \int_0^t \int_0^s \int_0^1 (s-r)^{\alpha-1} \sum_{i=1}^\infty E_{\alpha,\alpha}(-\lambda_i^\beta(s-r)^\alpha)\lambda_i^\beta \phi_i(x)$$

$$\times \phi_i(y)dW(r,y)ds$$

$$= \int_0^1 \sum_{i=1}^\infty tE_{\alpha,2}(-\lambda_i^\beta t^\alpha)\lambda_i^\beta \phi_i(x)\phi_i(y)u_0(y)dy$$

$$+ \int_0^1 \sum_{i=1}^\infty t^2 E_{\alpha,3}(-\lambda_i^\beta t^\alpha)\lambda_i^\beta \phi_i(x)\phi_i(y)v_0(y)dy$$

$$+ \int_0^t \int_0^1 (t-s)^\alpha \sum_{i=1}^\infty E_{\alpha,\alpha+1}(-\lambda_i^\beta(t-s)^\alpha)\lambda_i^\beta \phi_i(x)\phi_i(y)dW(s,y).$$

$$(3.89)$$

Substituting (3.88) and (3.89) into the left side of (3.87) gives

$$I_t^1[\partial_t^\alpha u(x,t)] + I_t^1[(-\Delta)^\beta u(x,t)]$$

$$= -\frac{t^{2-\alpha}}{\Gamma(3-\alpha)}v_0(x) + \int_0^1 \sum_{i=1}^\infty t^{2-\alpha}E_{\alpha,3-\alpha}(-\lambda_i^\beta t^\alpha)\phi_i(x)\phi_i(y)v_0(y)dy$$

$$+ \int_0^1 \sum_{i=1}^\infty t^2 E_{\alpha,3}(-\lambda_i^\beta t^\alpha)\lambda_i^\beta \phi_i(x)\phi_i(y)v_0(y)dy$$

$$+ \int_0^t \int_0^1 \sum_{i=1}^\infty E_{\alpha,1}(-\lambda_i^\beta(t-s)^\alpha)\phi_i(x)\phi_i(y)dW(s,y)$$

$$+ \int_0^t \int_0^1 (t-s)^\alpha \sum_{i=1}^\infty E_{\alpha,\alpha+1}(-\lambda_i^\beta(t-s)^\alpha)\lambda_i^\beta \phi_i(x)\phi_i(y)dW(s,y)$$

$$= \int_0^t \int_0^1 \sum_{i=1}^\infty \phi_i(x)\phi_i(y)dW(s,y) = \int_0^t \sum_{k=1}^\infty \sigma_i(s)\phi_i(x)d\xi_i(s)$$

$$= I_t^1\left[\frac{\partial^2 W(t,x)}{\partial t\partial x}\right].$$

$$(3.90)$$

The lemma is now proved. □

Similarly, the integral formulation of (3.79) is

$$u_m(x,t) = \int_0^1 \mathcal{T}_{\alpha,\beta}(x,y,t)u_0(y)\mathrm{d}y + \int_0^1 \mathcal{R}_{\alpha,\beta}(x,y,t)v_0(y)\mathrm{d}y$$
$$+ \int_0^1 \int_0^t \mathcal{S}_{\alpha,\beta}(x,y,t-s)\mathrm{d}W_m(y,s)$$

(3.91)

In what follows, the symbol C will denote a generic constant whose value may change from one line to another. First, we give the following stability estimates of the homogeneous problem which will play a key role in the error analysis of the finite element method (FEM) approximations.

Lemma 3.7 ([115]) *The solution $u(t)$ to the homogeneous problem of (3.78) satisfies, for $t > 0$,*

$$\|u(x,t)\|_{L^2(D,\dot{U}^p)}$$
$$\leq \begin{cases} Ct^{\frac{\alpha(q-p)}{2\beta}}\|u_0\|_{L^2(D,\dot{U}^q)} + Ct^{1-\frac{\alpha(p-r)}{2\beta}}\|v_0\|_{L^2(D,\dot{U}^r)}, \\ \qquad\qquad\qquad\qquad\qquad 0 \leq q,r < p \leq 2\beta, \\ Ct^{-\alpha}\|u_0\|_{L^2(D,\dot{U}^q)} + Ct^{1-\alpha}\|v_0\|_{L^2(D,\dot{U}^r)}, \qquad q,r > p, \end{cases}$$

(3.92)

and

$$\|\partial_t u(x,t)\|_{L^2(D,\dot{U}^p)}$$
$$\leq Ct^{-\alpha-\frac{\alpha(p-q)}{2\beta}}\|u_0\|_{L^2(D,\dot{U}^q)}$$
$$+ Ct^{1-\alpha-\frac{\alpha(p-r)}{2\beta}}\|v_0\|_{L^2(D,\dot{U}^r)}, \quad 0 \leq p \leq q, r < p+2\beta.$$

Under assumptions on $\{\sigma_i(t)\}$ and $\{\sigma_i^m(t)\}$, our first main result shows that the solution u_m of (3.79) indeed approximates u, the solution of (3.78).

Theorem 3.9 *Assume that $\{\sigma_i(t)\}$ and its derivative are uniformly bounded by*

$$|\sigma_i(t)| \leq \eta_i, \quad |\sigma_i'(t)| \leq \zeta_i \quad \forall t \in [0,T],$$

and that the coefficients $\{\sigma_i^m(t)\}$ are constructed such that

$$|\sigma_i(t) - \sigma_i^m(t)| \leq \vartheta_i^m, \quad |\sigma_i^m(t)| \leq \eta_i^m,$$
$$|(\sigma_i^m)'(t)| \leq \zeta_i^m \quad \forall t \in [0, T],$$

with positive sequences $\{\vartheta_i^m\}$ being arbitrarily chosen, and $\{\eta_i^m\}$ and $\{\zeta_i^m\}$ being related to $\{\vartheta_i^m \eta_i\}$ and $\{\zeta_i\}$. The series $(\{\vartheta_i^m\}, \{\eta_i^m\}, \{\zeta_i^m\})$ is required to rapidly decay to ensure the convergence of the series in (3.95). Let u_m and u be the solutions of (3.79) and (3.78), respectively. Then, for some constant $C > 0$ independent of Δt, any $0 < \varepsilon < \frac{1}{2}$ and any $1 < \alpha \leq \frac{3}{2}$,

$$E \|u(t) - u_m(t)\|^2 \leq C \sum_{i=1}^{\infty} \lambda_i^{-\frac{2\beta(\alpha-1)}{\alpha}} (\vartheta_i^m)^2$$

$$+ C(\Delta t)^2 \sum_{i=1}^{\infty} \lambda_i^{-\frac{2\beta(\alpha-1)}{\alpha}} (\zeta_i^m)^2$$

$$+ C t^{2\varepsilon} (\Delta t)^{2\alpha-1-2\varepsilon} \sum_{i=1}^{\infty} (\eta_i^m)^2, \quad t > 0.$$

$$(3.93)$$

For $\frac{3}{2} < \alpha < 2$,

$$E \|u(t) - u_m(t)\|^2 \leq C \sum_{i=1}^{\infty} \lambda_i^{-\frac{2\beta(\alpha-1)}{\alpha}} (\vartheta_i^m)^2$$

$$+ C(\Delta t)^2 \sum_{i=1}^{\infty} \lambda_i^{-\frac{2\beta(\alpha-1)}{\alpha}} (\zeta_i^m)^2 \qquad (3.94)$$

$$+ C t^{2\alpha-3} (\Delta t)^2 \sum_{i=1}^{\infty} (\eta_i^m)^2, \quad t > 0.$$

We also have the estimate, for any $\alpha \in (1, 2)$,

$$E \|u(t) - u_m(t)\|^2 \leq C \sum_{i=1}^{\infty} \lambda_i^{-\frac{2\beta(\alpha-1)}{\alpha}} (\vartheta_i^m)^2$$

$$+ C(\Delta t)^2 \sum_{i=1}^{\infty} \lambda_i^{-\frac{2\beta(\alpha-1)}{\alpha}} (\zeta_i^m)^2 \qquad (3.95)$$

$$+ C t^{2\alpha-2} \Delta t \sum_{i=1}^{\infty} (\eta_i^m)^2 + C(\Delta t)^2 \sum_{i=1}^{\infty} (\eta_i^m)^2, \quad t > 0,$$

provided that the infinite series are all convergent. So for $\alpha \in (1, \frac{3}{2}]$, there exists

$$\left(\mathrm{E} \, \|u(t) - u_m(t)\|^2 \right)^{\frac{1}{2}} \leq C \, (\Delta t)^{\max\{\frac{1}{2}, \alpha - \frac{1}{2} - \varepsilon\}}.$$

3.3.2.2   Galerkin finite element approximation

Next, we provide the Galerkin FEM scheme and derive the corresponding error estimates. For the sake of simplicity, we consider the same partition of $[0, 1]$: $0 = x_0 < x_2 \cdots < x_{N+1} = 1$ with $x_j = jh$ and $h = 1/(N + 1)$. Let V_h be the finite element subspace (with the order of the piecewise polynomial greater than or equal to 1) of $\dot{U}^\beta$.

On the space V_h we define the orthogonal L_2-projection P_h: $U \to V_h$ and the generalized Ritz projection R_h: $\dot{U}^\beta \to V_h$, respectively, by

$$(P_h \chi, v) = (\chi, v) \qquad \forall v \in V_h$$

and

$$\left((-\Delta)^{\frac{\beta}{2}} R_h \chi, (-\Delta)^{\frac{\beta}{2}} v \right) = \left((-\Delta)^{\frac{\beta}{2}} \chi, (-\Delta)^{\frac{\beta}{2}} v \right) \qquad \forall v \in V_h.$$

The projection R_h of χ is unique since $\chi \in \dot{U}^\beta$, and it equals to zero on the boundary. Since $\dot{U}^\beta$ is equivalent to the fractional Sobolev space U_0^β for $q \in [0, 1]$, and $\|\psi\|_{U^{1+s}} \leq C\|\Delta\psi\|_{\dot{U}^{-1+s}} = C\|\psi\|_{\dot{U}^{1+s}}$ for $0 \leq s < \beta$, $\beta \in \left(\frac{1}{2}, 1 \right)$ being established in [174], we have the following error estimates for $P_h \chi$ and $R_h \chi$.

Lemma 3.8 *The operators P_h and R_h satisfy*

$$\|P_h \psi - \psi\| + h^\beta \left\| (-\Delta)^{\frac{\beta}{2}} (P_h \psi - \psi) \right\| \leq Ch^q \|\psi\|_q \qquad \text{for}$$
$$\psi \in \dot{U}^q, \quad q \in [\beta, 2\beta],$$

and

$$\|R_h \psi - \psi\| + h^\beta \left\| (-\Delta)^{\frac{\beta}{2}} (R_h \psi - \psi) \right\| \leq Ch^q \|\psi\|_q \qquad \text{for}$$
$$\psi \in \dot{U}^q, \quad q \in [\beta, 2\beta].$$

Upon introducing the discrete fractional Laplacian $(-\Delta_h)^\beta$: $V_h \to V_h$ defined by

$$\left((-\Delta_h)^\beta \psi, \chi \right) = \left((-\Delta)^{\frac{\beta}{2}} \psi, (-\Delta)^{\frac{\beta}{2}} \chi \right) \qquad \forall \psi, \, \chi \in V_h, \qquad (3.96)$$

we can write the spatial FEM approximation of (3.79) as

$$\partial_t^\alpha u_m^h(t) = -(-\Delta_h)^\beta u_m^h(t) + P_h \frac{\partial^2 W_m(x,t)}{\partial t \partial x},$$

$$0 < t \le T, \ with \ u_m^h(0) = u_0^h, \ \partial_t u_m^h(0) = v_0^h, \tag{3.97}$$

where $u_0^h = P_h u_0$, $v_0^h = P_h v_0$, or $v_0^h = R_h v_0$.

Now we give a representation of the solution of (3.79) using the eigenvalues and eigenfunctions $\{\lambda_i^{h,\beta}\}_{i=1}^N$ and $\{\psi_i^h\}_{i=1}^N$ of the discrete fractional Laplacian $(-\Delta_h)^\beta$. Since we know that the operator $(-\Delta_h)^\beta$ is symmetrical, $\{\psi_i^h\}_{i=1}^N$ is therefore orthogonal. Take $\{\psi_i^h\}_{i=1}^N$ as the orthonormal bases in V_h and define the discrete analogues of (3.81)-(3.83) by

$$\mathcal{T}_{\alpha,\beta}^h(x,y,t) = \sum_{i=1}^N E_{\alpha,1}(-\lambda_i^{h,\beta} t^\alpha) \phi_i^h(x) \phi_i^h(y), \tag{3.98}$$

$$\mathcal{R}_{\alpha,\beta}^h(x,y,t) = \sum_{i-1}^N t E_{\alpha,2}(-\lambda_i^{h,\beta} t^\alpha) \phi_i^h(x) \phi_i^h(y), \tag{3.99}$$

and

$$\mathcal{S}_{\alpha,\beta}^h(x,y,t) = \sum_{i=1}^N t^{\alpha-1} E_{\alpha,\alpha}(-\lambda_i^{h,\beta} t^\alpha) \phi_i^h(x) \phi_i^h(y). \tag{3.100}$$

Then the solution u_m^h of the discrete problem (3.97) can be expressed by

$$u_m^h(x,t) = \int_0^1 \mathcal{T}_{\alpha,\beta}^h(x,y,t) u_0^h(y) dy + \int_0^1 \mathcal{R}_{\alpha,\beta}^h(x,y,t) v_0^h(y) dy$$

$$+ \int_0^1 \int_0^t \mathcal{S}_{\alpha,\beta}^h(x,y,t-s) P_h dW_m(y,s).$$

$$\tag{3.101}$$

Also, on the finite element space V_h, we introduce the discrete norm $\|\cdot\|_{p,h}$ for any $p \in \mathbb{R}$ defined by

$$\|\psi\|_{p,h}^2 = \sum_{i=1}^N \left(\lambda_i^{h,\beta}\right)^{\frac{p}{\beta}} \left(\chi, \phi_i^h\right)^2, \qquad \chi \in V_h. \tag{3.102}$$

Clearly, the norm $\| \cdot \|_{p,h}$ is well defined for all real p. By the definition of the discrete fractional Laplacian $(-\Delta_h)^\beta$, we have $\|\psi\|_{p,h} = \|\psi\|_p$ for $p = 0,\ \beta$ and for all $\psi \in V_h$. So there is no confusion in using $\|\psi\|_p$ instead of $\|\psi\|_{p,h}$ for $p = 0,\ \beta$ and for all $\psi \in V_h$. Further, we need the following inverse inequality.

Lemma 3.9 *For any $l > s$, there exists a constant C independent of h such that*

$$\|\chi\|_{l,h} \leq h^{s-l}\|\chi\|_{s,h} \quad \forall v \in V_h. \tag{3.103}$$

Proof: *For all $\chi \in V_h$, the inverse inequality $\|\chi\|_\beta \leq Ch^\beta\|\chi\|$ holds [25]. By the definition of $(-\Delta_h)^\beta$, there exists $\max_{1 \leq i \leq N} \lambda_i^{h,\beta} \leq Ch^{-2\beta}$. Thus, for the norm $\| \cdot \|_{p,h}$ defined in (3.102), there holds, for any real $l > s$,*

$$\|\chi\|_{l,h}^2 \leq C \max_{1 \leq i \leq N} \left(\lambda_i^{h,\beta}\right)^{\frac{l-s}{\beta}} \sum_{i=1}^{N} \left(\lambda_i^{h,\beta}\right)^{\frac{s}{\beta}} \left(\chi, \phi_i^h\right)^2 \leq h^{2(s-l)}\|\chi\|_{s,h}^2.$$

$\square$

The following estimates are crucial for the error analysis in what follows.

Lemma 3.10 *Let $\mathcal{T}_{\alpha,\beta}^h(t,x,y)$ be defined by (3.98) and $u_0^h \in V_h$. Then, for all $t > 0$,*

$$\left\| \int_0^1 \mathcal{T}_{\alpha,\beta}^h(t,x,y)u_0^h(y)dy \right\|_{p,h}$$

$$\leq \begin{cases} Ct^{\frac{\alpha(q-p)}{2\beta}} \|u_0^h\|_{q,h}, & 0 \leq q \leq p \leq 2\beta, \\ \\ Ct^{-\alpha} \|u_0^h\|_{q,h}, & q > p. \end{cases}$$

Lemma 3.11 *Let $\mathcal{R}_{\alpha,\beta}^h(t,x,y)$ be defined by (3.99) and $v_0^h \in V_h$. Then, for all $t > 0$,*

$$\left\| \int_0^1 \mathcal{R}_{\alpha,\beta}^h(t,x,y)v_0^h(y)dy \right\|_{p,h}$$

$$\leq \begin{cases} Ct^{1-\frac{\alpha(p-q)}{2\beta}} \|v_0^h\|_{q,h}, & 0 \leq q \leq p \leq 2\beta, \\ \\ Ct^{1-\alpha} \|v_0^h\|_{q,h}, & q > p. \end{cases}$$

The proofs of Lemmas 3.10 and 3.11 are discrete analogues of those formulated in (3.92), so here we omit them.

Lemma 3.12 *Let $\mathcal{S}_{\alpha,\beta}^h(t, x, y)$ be defined by (3.100) and $\psi \in V_h$. Then, for all $t > 0$,*

$$\left| \int_0^1 \mathcal{S}_{\alpha,\beta}^h(t, x, y)\psi(y)dy \right|_{p,h}$$

$$\leq \begin{cases} Ct^{-1+\alpha+\frac{\alpha(q-p)}{2\beta}} |\psi|_{q,h}, & p - 2\beta \leq q \leq p, \\ Ct^{-1}|\psi|_{q,h}, & q > p. \end{cases}$$

Proof: By Lemmas 3.4 and 3.6, we obtain that

$$\left| \int_0^1 \mathcal{S}_{\alpha,\beta}^h(t, x, y)\psi(y)dy \right|_{p,h}^2$$

$$= \sum_{i=1}^N (\lambda_i^{h,\beta})^{\frac{p}{\beta}} t^{2\alpha-2} |E_{\alpha,\alpha}(-\lambda_i^{h,\beta} t^\alpha)|^2 (\psi, \phi_i^h)^2 \qquad (3.104)$$

$$\leq t^{\frac{\alpha(q-p)}{\beta}-2+2\alpha} \max_i \frac{C\left(\lambda_i^{h,\beta} t^\alpha\right)^{\frac{p-q}{\beta}}}{(1+\lambda_i^{h,\beta} t^\alpha)^2} \sum_{i=1}^N (\lambda_i^{h,\beta})^{\frac{q}{\beta}}(\psi, \phi_i^h)^2$$

$$\leq Ct^{\frac{\alpha(q-p)}{\beta}-2+2\alpha}|\psi|_{q,h}^2,$$

where we have used $\dfrac{(\lambda_i^{h,\beta} t^\alpha)^{\frac{p-q}{\beta}}}{(1+\lambda_i^{h,\beta} t^\alpha)^2} \leq C$ for $p - 2\beta \leq q \leq p$.

For $q > p$, since $\{\lambda_i^{h,\beta}\}$ are bounded away from zero independent of the mesh size h, we deduce from Lemmas 3.4 and 3.6 that

$$\left| \int_0^1 \mathcal{S}_{\alpha,\beta}^h(t, x, y)\psi(y)dy \right|_{p,h}^2$$

$$\leq t^{2\alpha-2} \max_i \frac{C}{(1+\lambda_i^{h,\beta} t^\alpha)^2 (\lambda_i^{h,\beta})^{\frac{q-p}{\beta}}} \sum_{i=1}^N (\lambda_i^{h,\beta})^{\frac{q}{\beta}}(\psi, \phi_i^h)^2 \qquad (3.105)$$

$$\leq Ct^{-2}|\psi|_{q,h}^2.$$

Thus, we complete the proof of Lemma 3.12. $\qquad\qquad\square$

We have the following theorem for the regularity of the solution of (3.79).

Theorem 3.10 *Let u_m be the solution of (3.79). Assume that $\{\sigma_k^m(t)\}$ are uniformly bounded by $|\sigma_k^m(t)| \leq \eta_k^m \; \forall t \in [0, T]$. Further assume that $u_0 \in L^2(D; \dot{U}^q)$, $v_0 \in L^2(D; \dot{U}^r)$, $q, r \in [0, 2\beta]$. Then, for some constant $C > 0$ independent of Δt, we have, for all $t > 0$,*

$$
\begin{aligned}
\mathbf{E}\|u_m(t)\|_{2\beta}^2 &\leq Ct^{\frac{\alpha(q-2\beta)}{\beta}}\mathbf{E}\|u_0\|_q^2 + Ct^{2-\frac{\alpha(2\beta-r)}{\beta}}\mathbf{E}\|v_0\|_r^2 \\
&\quad + C\sum_{i=1}^{\infty}\lambda_i^{\frac{2\beta}{\alpha}}(\eta_i^m)^2
\end{aligned}
\tag{3.106}
$$

and

$$
\begin{aligned}
\mathbf{E}\|\partial_t^\alpha u_m(t)\|^2 &\leq Ct^{\frac{\alpha(q-2\beta)}{\beta}}\mathbf{E}\|u_0\|_q^2 + Ct^{2-\frac{\alpha(2\beta-r)}{\beta}}\mathbf{E}\|v_0\|_r^2 \\
&\quad + C\sum_{i=1}^{\infty}\lambda_i^{\frac{2\beta}{\alpha}}(\eta_i^m)^2 + C\frac{1}{\Delta t}\sum_{i=1}^{\infty}(\eta_i^m)^2,
\end{aligned}
\tag{3.107}
$$

provided that the infinite series are all convergent.

Proof: It follows from Lemmas 3.6 and 3.7 that

$$
\begin{aligned}
&\mathbf{E}\|u_m(t)\|_{2\beta}^2 \\
&\leq Ct^{\frac{\alpha(q-2\beta)}{\beta}}\mathbf{E}\|u_0\|_q^2 + Ct^{2-\frac{\alpha(2\beta-r)}{\beta}}\mathbf{E}\|v_0\|_r^2 \\
&\quad + C\mathbf{E}\left\|\int_0^t (t-s)^{\alpha-1}\sum_{i=1}^{\infty}\int_0^1 E_{\alpha,\alpha}(-\lambda_i^\beta(t-s)^\alpha)e_i(y) \right. \\
&\quad \left. \times\, e_i(x)dW_m(s,y)\right\|_{2\beta}^2 .
\end{aligned}
\tag{3.108}
$$

Let

$$
\begin{aligned}
\mathcal{F}_1 \\
= C\mathbf{E}\left\|\int_0^t (t-s)^{\alpha-1}\sum_{i=1}^{\infty}\int_0^1 E_{\alpha,\alpha}(-\lambda_i^\beta(t-s)^\alpha)\right. \\
\left. e_i(y)e_i(x)dW_m(s,y)\right\|_{2\beta}^2 .
\end{aligned}
$$

Without loss of generality, we assume that there exists a positive integer I_t such that $t = t_{I_t+1}$. Since $\{e_i\}_{i=1}^{\infty}$ is an orthonormal basis in $L^2(0,1)$, and the Brownian motion has independent increments, then by Hölder's inequality and the boundedness assumption on $\sigma_i^m(t)$, we have

$$\mathcal{F}_1 = C\mathbf{E}\sum_{i=1}^{\infty}\lambda_i^{2\beta}\left(\int_0^t(t-s)^{\alpha-1}\int_0^1 E_{\alpha,\alpha}(-\lambda_i^{\beta}(t-s)^{\alpha})\right.$$

$$\left. \times\, e_i(y)dW_m(s,y)\right)^2$$

$$= C\sum_{i=1}^{\infty}\lambda_i^{2\beta}\sum_{k=1}^{I_t}\frac{\mathbf{E}(\xi_i(t_{k+1})-\xi_i(k_i))^2}{(\Delta t)^2}$$

$$\times\left(\int_{t_k}^{t_{k+1}}(t-s)^{\alpha-1}E_{\alpha,\alpha}(-\lambda_i^{\beta}(t-s)^{\alpha})\sigma_i^m(s)ds\right)^2$$

$$\leq C\sum_{i=1}^{\infty}\lambda_i^{2\beta}\sum_{k=1}^{I_t}\int_{t_k}^{t_{k+1}}((t-s)^{\alpha-1}E_{\alpha,\alpha}(-\lambda_i^{\beta}(t-s)^{\alpha}))^2(\sigma_i^m(s))^2ds$$

$$\leq C\sum_{i=1}^{\infty}\lambda_i^{2\beta}(\mu_i^m)^2\int_0^t\lambda_i^{-\frac{2\beta(\alpha-1)}{\alpha}}\left|\frac{(\lambda_i^{\beta}(t-s)^{\alpha})^{\frac{\alpha-1}{\alpha}}}{1+\lambda_i^{\beta}(t-s)^{\alpha}}\right|^2ds$$

$$\leq C\sum_{i=1}^{\infty}\lambda_i^{\frac{2\beta}{\alpha}}(\mu_i^m)^2,$$

$$(3.109)$$

where we have used $\dfrac{(\lambda_i^{\beta}(t-s)^{\alpha})^{\frac{\alpha-1}{\alpha}}}{1+\lambda_i^{\beta}(t-s)^{\alpha}} \leq C$.

Thus, the conclusion (3.106) follows immediately from (3.108) and (3.109).

Now we show (3.107). Note that $\{e_i\}_{i=1}^{\infty}$ is an orthonormal basis in $L^2(0,1)$, and for any $t \in [0,T]$, there exists $k \in \{1,2,\ldots,I\}$ such that $t \in [t_k, t_{k+1}]$. By (3.108) and (3.109), we deduce that

$$\mathbf{E}\|\partial_t^\alpha u_m(t)\|^2 = \mathbf{E}\int_0^1 (\partial_t^\alpha u_m(t,x))^2 dx$$

$$\leq Ct^{\frac{\alpha(q-2\beta)}{\beta}}\mathbf{E}\|u_0\|_q^2 + Ct^{2-\frac{\alpha(2\beta-r)}{\beta}}\mathbf{E}\|v_0\|_r^2 + C\sum_{i=1}^{\infty}\lambda_i^{\frac{2\beta}{\alpha}}(\mu_k^m)^2$$

$$+ C\sum_{i=1}^{\infty}(\sigma_i^m(t))^2 \frac{1}{(\Delta t)^2}\mathbf{E}(\xi_i(t_{k+1}) - \xi_i(t_k))^2$$

$$\leq Ct^{\frac{\alpha(q-2\beta)}{\beta}}\mathbf{E}\|u_0\|_q^2 + Ct^{2-\frac{\alpha(2\beta-r)}{\beta}}\mathbf{E}\|v_0\|_r^2 + C\sum_{i=1}^{\infty}\lambda_i^{\frac{2\beta}{\alpha}}(\eta_i^m)^2$$

$$+ C\frac{1}{\Delta t}\sum_{i=1}^{\infty}(\eta_i^m)^2.$$

$$(3.110)$$

□

The proof is completed.

First, we establish error estimates for the homogeneous problem (3.79) with initial data $u_0 \in \dot{U}^q, q \in [\beta, 2\beta], v_0 = 0$.

Theorem 3.11 *Let u_m be the solution of the homogeneous problem (3.79) with $u_0 \in \dot{U}^q, q \in [\beta, 2\beta], v_0 = 0$, and let u_m^h be the solution of the homogeneous problem (3.97) with $u_0^h = P_h u_0, v_0^h = 0$. Then, with $\ell_h = |\ln h|$,*

$$\|u_m^h(t) - u_m(t)\| + h^\beta\|(-\Delta)^{\frac{\beta}{2}}(u_m^h(t) - u_m(t))\|$$

$$\leq C\ell_h h^{2\beta} t^{\frac{\alpha q}{2\beta} - \alpha}\|u_0\|_q \quad \forall t \in (0, T].$$

Proof: For $u_0 \in \dot{U}^q, q \in [\beta, 2\beta], v_0 = 0$, we split the error $u_m^h - u_m$ into two terms as

$$u_m^h - u_m = (u_m^h - P_h u_m) + (P_h u_m - u_m) := err1 + err2.$$

It follows from Lemmas 3.7 and 3.8 that

$$\|err2(t)\| + h^\beta\|(-\Delta)^{\frac{\beta}{2}}err2(t)\| \leq Ch^{2\beta}\|u_m(t)\|_{2\beta}$$

$$\leq Ch^{2\beta}t^{\frac{\alpha q}{2\beta} - \alpha}\|u_0\|_q.$$

$$(3.111)$$

From the definitions of P_h and R_h, it is easy to check that $(-\Delta_h)^\beta R_h = P_h(-\Delta)^\beta$. Then we can see that $err1$ satisfies

$$\partial_t^\alpha err1 + (-\Delta_h)^\beta err1 = (-\Delta_h)^\beta (R_h u_m - P_h u_m),$$

with $err1(0) = \partial_t err1(0) = 0$. Then (3.101) implies that

$$err1(t,x) = \int_0^t \int_0^1 S_{\alpha,\beta}^h(t-s,x,y)(-\Delta_h)^\beta(R_h u_m(s,y)$$

$$- P_h u_m(s,y))dy ds.$$

For any $0 < \varepsilon < 2\beta$, by (3.103) and Lemmas 3.7, 3.8 and 3.11, we obtain that

$$\|err1(t)\| \le \int_0^t \left\| \int_0^1 S_{\alpha,\beta}^h(t-s,x,y)(-\Delta_h)^\beta(R_h u_m(s,y) \right.$$

$$\left. - P_h u_m(s,y))dy \right\| ds$$

$$\le C \int_0^t (t-s)^{\frac{\alpha\varepsilon}{2\beta}-1} \|(-\Delta_h)^\beta(R_h u_m - P_h u_m)(s)\|_{\varepsilon-2\beta,h} ds$$

$$= C \int_0^t (t-s)^{\frac{\alpha\varepsilon}{2\beta}-1} \|(R_h u_m - P_h u_m)(s)\|_{\varepsilon,h} ds$$

$$\le Ch^{-\varepsilon} \int_0^t (t-s)^{\frac{\alpha\varepsilon}{2\beta}-1} \|(R_h u_m - P_h u_m)(s)\| ds$$

$$\le Ch^{2\beta-\varepsilon} \int_0^t (t-s)^{\frac{\alpha\varepsilon}{2\beta}-1} \|u_m(s)\|_{2\beta} ds$$

$$\le Ch^{2\beta-\varepsilon} \int_0^t (t-s)^{\frac{\alpha\varepsilon}{2\beta}-1} s^{\frac{\alpha q}{2\beta}-\alpha} \|u_0\|_q ds$$

$$\le CB\left(\frac{\alpha\varepsilon}{2\beta}, 1 + \frac{\alpha q}{2\beta} - \alpha\right) h^{2\beta-\varepsilon} t^{\frac{\alpha q}{2\beta}-\alpha} \|u_0\|_q$$

$$\le C\varepsilon^{-1} h^{2\beta-\varepsilon} t^{\frac{\alpha q}{2\beta}-\alpha} \|u_0\|_q.$$

$$(3.112)$$

In a similar way, we have

$$h^\beta \|(-\Delta)^{\frac{\beta}{2}} err1(t)\|$$

$$\leq Ch^\beta \int_0^t (t-s)^{\frac{\alpha\varepsilon}{2\beta}-1} \|(-\Delta_h)^\beta (R_h u_m - P_h u_m)(s)\|_{\varepsilon-\beta,h} ds$$

$$= Ch^\beta \int_0^t (t-s)^{\frac{\alpha\varepsilon}{2\beta}-1} \|(R_h u_m - P_h u_m)(s)\|_{\varepsilon+\beta,h} ds$$

$$\leq Ch^{2\beta-\varepsilon} \int_0^t (t-s)^{\frac{\alpha\varepsilon}{2\beta}-1} \|(R_h u_m - P_h u_m)(s)\|_\beta ds$$

$$\leq Ch^{2\beta-\varepsilon} \int_0^t (t-s)^{\frac{\alpha\varepsilon}{2\beta}-1} \|u_m(s)\|_{2\beta} ds$$

$$\leq Ch^{2\beta-\varepsilon} \int_0^t (t-s)^{\frac{\alpha\varepsilon}{2\beta}-1} s^{\frac{\alpha q}{2\beta}-\alpha} \|u_0\|_q ds$$

$$\leq CB\left(\frac{\alpha\varepsilon}{2\beta}, 1 + \frac{\alpha q}{2\beta} - \alpha\right) h^{2\beta-\varepsilon} t^{\frac{\alpha q}{2\beta}-\alpha} \|u_0\|_q$$

$$\leq C\varepsilon^{-1} h^{2\beta-\varepsilon} t^{\frac{\alpha q}{2\beta}-\alpha} \|u_0\|_q.$$

$$(3.113)$$

The last inequalities in (3.112) and (3.113) follow from the fact that $B(\frac{\alpha\varepsilon}{2\beta}, 1 + \frac{\alpha q}{2\beta} - \alpha) = \frac{\Gamma(\frac{\alpha\varepsilon}{2\beta})\Gamma(1+\frac{\alpha q}{2\beta}-\alpha)}{\Gamma(\frac{\alpha\varepsilon}{2\beta}+1+\frac{\alpha q}{2\beta}-\alpha)}$ and $\Gamma(\frac{\alpha\varepsilon}{2\beta}) \sim \frac{2\beta}{\alpha\varepsilon}$ as $\varepsilon \to 0^+$, e.g., by means of Laurent expansion of the Gamma function. Then the desired assertion follows by choosing $\varepsilon = 1/\ell_h$ and the triangle inequality. $\qquad\square$

Next we state an error estimate for the homogeneous problem (3.79) with initial data $u_0 = 0$, $v_0 \in \dot{U}^{2\beta}$.

Theorem 3.12 *Let u_m be the solution of the homogeneous problem (3.79) with $u_0 = 0, v_0 \in \dot{U}^{2\beta}$, and u_m^h be the solution of the homogeneous problem (3.97) with $u_0^h = 0$, $v_0^h = R_h v_0$. Then*

$$\|u_m^h(t) - u_m(t)\| + h^\beta \|(-\Delta)^{\frac{\beta}{2}}(u_m^h(t) - u_m(t))\|$$
$$\leq Ch^{2\beta} t \|v_0\|_{2\beta} \qquad \forall t \in (0, T].$$

Proof: For $u_0 = 0, v_0 \in \dot{U}^{2\beta}$, we split the error $u_m^h - u_m$ into two terms as

$$u_m^h - u_m = (u_m^h - R_h u_m) + (R_h u_m - u_m) := \bar{err}1 + \bar{err}2.$$

By Lemmas 3.7 and 3.8, we have for any $t > 0$

$$\|\bar{err}2(t)\| + h^\beta \|(-\Delta)^{\frac{\beta}{2}} \bar{err}2(t)\| \leq Ch^{2\beta} \|u_m(t)\|_{2\beta} \leq Ch^{2\beta} t \|v_0\|_{2\beta}.$$
$$(3.114)$$

Using the identity $(-\Delta_h)^\beta R_h = P_h(-\Delta)^\beta$, we note that $\bar{err}1$ satisfies

$$\partial_t^\alpha \bar{err}1(t) + (-\Delta_h)^\beta \bar{err}1(t) = -P_h \partial_t^\alpha \bar{err}2(t),$$

where $\bar{err}1(0) = \partial_t \bar{err}1(0) = 0$. By (3.101), we obtain that

$$\bar{err}1(t,x) = -\int_0^t \int_0^1 S_{\alpha,\beta}^h(t-s,x,y) P_h \partial_s^\alpha \bar{err}2(s,y) dy ds.$$

Applying Lemmas 3.7, 3.8 and 3.12, then we deduce that for $p = 0, \beta$,

$$\|\bar{err}1(t)\|_p \leq \int_0^t \left\| \int_0^1 S_{\alpha,\beta}^h(t-s,x,y) P_h \partial_s^\alpha \bar{err}2(s,y) dy \right\|_p ds$$

$$\leq C \int_0^t (t-s)^{\alpha-1} \|\partial_s^\alpha \bar{err}2(s)\|_p ds$$

$$= C \int_0^t (t-s)^{\alpha-1} \|(R_h \partial_s^\alpha u_m - \partial_s^\alpha u_m)(s)\|_p ds$$

$$\leq Ch^{2\beta-p} \int_0^t (t-s)^{\alpha-1} \|\partial_s^\alpha u_m(s)\|_{2\beta} ds$$

$$\leq Ch^{2\beta-p} \int_0^t (t-s)^{\alpha-1} s^{1-\alpha} \|v_0\|_{2\beta} ds$$

$$\leq Ch^{2\beta-p} B(\alpha, 2-\alpha) t \|v_0\|_{2\beta}.$$
$$(3.115)$$

Thus the conclusion follows immediately by the triangle inequality and (3.114)-(3.115). □

Finally we show an error estimate for the homogeneous problem (3.79) with initial data $u_0 = 0$, $v_0 \in H$.

Theorem 3.13 *Let u_m be the solution of the homogeneous problem (3.79) with $u_0 = 0, v_0 \in H$, and u_m^h be the solution of the homogeneous problem (3.97) with $u_0^h = 0$, $v_0^h = P_h v_0$. Then with $\ell_h = |\ln h|$,*

$$\|u_m^h(t) - u_m(t)\| + h^\beta \|(-\Delta)^{\frac{\beta}{2}}(u_m^h(t) - u_m(t))\|$$
$$\leq C\ell_h h^{2\beta} t^{1-\alpha} \|v_0\| \quad \forall t \in (0,T].$$
$$(3.116)$$

Proof: For $u_0 = 0, v_0 \in H$, we split the error $u_m^h - u_m$ into two terms:

$$u_m^h - u_m = (u_m^h - P_h u_m) + (P_h u_m - u_m) := e\hat{e}r1 + e\hat{e}r2.$$

It follows from Lemmas 3.7 and 3.8 that

$$\|e\hat{e}r2(t)\| + h^\beta \|(-\Delta)^{\frac{\beta}{2}} e\hat{e}r2(t)\| \le Ch^{2\beta} \|u_m(t)\|_{2\beta} \le Ch^{2\beta} t^{1-\alpha} \|v_0\|.$$

Using the identity $(-\Delta_h)^\beta R_h = P_h(-\Delta)^\beta$, we see that $e\hat{e}r1$ satisfies

$$\partial_t^\alpha e\hat{e}r1 + (-\Delta_h)^\beta e\hat{e}r1 = (-\Delta_h)^\beta (R_h u_m - P_h u_m)$$

with $e\hat{e}r1(0) = \partial_t e\hat{e}r1(0) = 0$. Then (3.101) implies that

$$e\hat{e}r1(t,x) = \int_0^t \int_0^1 S_{\alpha,\beta}^h(t-s,x,y)(-\Delta_h)^\beta (R_h u_m(s,y)$$
$$- P_h u_m(s,y))dyds.$$

For any $0 < \varepsilon < 2\beta$, by (3.103) and Lemmas 3.7, 3.8 and 3.12, we obtain that for $p = 0, \beta$,

$$\|e\hat{e}r1(t)\|_p \le C \int_0^t (t-s)^{\frac{\alpha\varepsilon}{2\beta}-1} \|(-\Delta_h)^\beta (R_h u_m - P_h u_m)$$
$$\times (s)\|_{\varepsilon-2\beta+p,h} ds$$
$$= C \int_0^t (t-s)^{\frac{\alpha\varepsilon}{2\beta}-1} \|(R_h u_m - P_h u_m)(s)\|_{\varepsilon+p,h} ds$$
$$\le Ch^{-\varepsilon} \int_0^t (t-s)^{\frac{\alpha\varepsilon}{2\beta}-1} \|(R_h u_m - P_h u_m)(s)\|_p ds$$
$$\le Ch^{2\beta-p-\varepsilon} \int_0^t (t-s)^{\frac{\alpha\varepsilon}{2\beta}-1} \|u_m(s)\|_{2\beta} ds$$
$$\le Ch^{2\beta-p-\varepsilon} \int_0^t (t-s)^{\frac{\alpha\varepsilon}{2\beta}-1} s^{1-\alpha} \|v_0\| ds$$
$$\le CB\left(\frac{\alpha\varepsilon}{2\beta}, 2-\alpha\right) h^{2\beta-p-\varepsilon} t^{1-\alpha} \|v_0\|$$
$$\le C\varepsilon^{-1} h^{2\beta-p-\varepsilon} t^{1-\alpha} \|v_0\|.$$

$$(3.117)$$

The last inequality follows from the fact $B(\frac{\alpha\varepsilon}{2\beta}, 2-\alpha) = \frac{\Gamma(\frac{\alpha\varepsilon}{2\beta})\Gamma(2-\alpha)}{\Gamma(\frac{\alpha\varepsilon}{2\beta}+2-\alpha)}$ and $\Gamma(\frac{\alpha\varepsilon}{2\beta}) \sim \frac{2\beta}{\alpha\varepsilon}$ as $\varepsilon \to 0^+$. Then the desired assertion follows by choosing $\varepsilon = 1/\ell_h$ and the triangle inequality. $\qquad\square$

Remark 3.2 *Let u_m be the solution of the homogeneous problem (3.79) with $u_0 = 0, v_0 \in \dot{U}^r, r \in [0, 2\beta]$, and u_m^h be the solution of the homogeneous problem (3.97) with $u_0^h = 0, v_0^h = P_h v_0$. Then with $\ell_h = |\ln h|$,*

$$\|u_m^h(t) - u_m(t)\| + h^\beta \|(-\Delta)^{\frac{\beta}{2}} (u_m^h(t) - u_m(t))\|$$
$$\leq C\ell_h h^{2\beta} t^{1-\alpha+\frac{\alpha r}{2\beta}} |v_0|_r \quad \forall t \in (0, T]. \tag{3.118}$$

3.3.2.3 Error estimates for the nonhomogeneous problem

First, we derive an error estimate for the nonhomogeneous problem (3.79) with initial data $u_0 = v_0 = 0$.

Theorem 3.14 *Let u_m and u_m^h be the solutions of (3.79) and (3.97) with $u_0 = v_0 = 0$, respectively. Assume that $\{\sigma_i^m(t)\}$ are uniformly bounded by $|\sigma_i^m(t)| \leq \eta_i^m \ \forall t \in [0, T]$. Then with $\ell_h = |\ln h|$,*

$$\mathbf{E}\|u_m^h(t) - u_m(t)\|^2 + h^{2\beta} \mathbf{E}\|(-\Delta)^{\frac{\beta}{2}} (u_m^h(t) - u_m(t))\|^2$$
$$\leq C\ell_h h^{4\beta} \frac{1}{\Delta t} \sum_{k=1}^{\infty} \lambda_i^{\frac{\beta(\alpha+1)}{\alpha}} (\eta_i^m)^2 \quad \forall t \in [0, T],$$

provided that the infinite series are convergent, where C is a positive constant independent of Δt and h.

Proof: We split the error $u_m^h - u_m$ into

$$u_m^h - u_m = (u_m^h - P_h u_m) + (P_h u_m - u_m) := v + \rho.$$

By Theorem 3.8 we have

$$\mathbf{E}\|\rho(t)\|^2 + h^{2\beta} \mathbf{E}\|(-\Delta)^{\frac{\beta}{2}} \rho(t)\|^2 \leq Ch^{4\beta} \mathbf{E}|u_m(t)|_{2\beta}^2. \tag{3.119}$$

Note that $u_0 = v_0 = 0$, hence it follows from (3.108) and (3.109) that

$$\mathbf{E}|u_m(t)|_{2\beta}^2 = \mathbf{E}\left| \int_0^t \int_0^1 S_{\alpha,\beta}(t - s, x, y) dW_m(s, y) \right|_{2\beta}^2 \tag{3.120}$$
$$\leq Ct^\alpha \frac{1}{\Delta t} \sum_{k=1}^{\infty} \lambda_i^{\frac{\beta(\alpha+1)}{\alpha}} (\eta_i^m)^2.$$

Therefore

$$\mathbf{E}\|\rho(t)\|^2 + h^{2\beta} \mathbf{E}\|(-\Delta)^{\frac{\beta}{2}} \rho(t)\|^2 \leq Ct^\alpha h^{4\beta} \frac{1}{\Delta t} \sum_{k=1}^{\infty} \lambda_i^{\frac{\beta(\alpha+1)}{\alpha}} (\eta_i^m)^2.$$
$$\tag{3.121}$$

Moreover, we consider the equation

$$\partial_t^\alpha v + (-\Delta_h)^\beta v = (-\Delta_h)^\beta (R_h u_m - P_h u_m) \text{ with } v(0) = \partial_t v(0) = 0.$$

Then it follows from (3.101) that

$$v(t,x) = \int_0^t \int_0^1 S_{\alpha,\beta}^h(t-s,x,y)(-\Delta_h)^\beta(R_h u_m(s,y)$$
$$- P_h u_m(s,y))dyds.$$

For any $0 < \varepsilon < 2\beta$, by the similar argument of (3.117) and using Hölder's inequality, we deduce that for $p = 0, \beta$,

$$\mathbf{E}|v(t)|_p^2 \le C\mathbf{E}\left(\int_0^t (t-s)^{\frac{\alpha\varepsilon}{2\beta}-1}|(-\Delta_h)^\beta(R_h u_m - P_h u_m)\right.$$
$$\left.(s)|_{\varepsilon-2\beta+p,h}ds\right)^2$$
$$= C\mathbf{E}\left(\int_0^t (t-s)^{\frac{\alpha\varepsilon}{2\beta}-1}|(R_h u_m - P_h u_m)(s)|_{\varepsilon+p,h}ds\right)^2$$
$$\le Ch^{-2\varepsilon}\mathbf{E}\left(\int_0^t (t-s)^{\frac{\alpha\varepsilon}{2\beta}-1}|(R_h u_m - P_h u_m)(s)|_p ds\right)^2$$
$$\le Ch^{4\beta-2p-2\varepsilon}\mathbf{E}\left(\int_0^t (t-s)^{\frac{\alpha\varepsilon}{2\beta}-1}|u_m(s)|_{2\beta}ds\right)^2$$
$$\le Ch^{4\beta-2p-2\varepsilon}\int_0^t (t-s)^{\frac{\alpha\varepsilon}{2\beta}-1}ds\int_0^t (t-s)^{\frac{\alpha\varepsilon}{2\beta}-1}\mathbf{E}|u_m(s)|_{2\beta}^2 ds$$
$$\le Ch^{4\beta-2p-2\varepsilon}t^{\frac{\alpha\varepsilon}{2\beta}}\int_0^t (t-s)^{\frac{\alpha\varepsilon}{2\beta}-1}\mathbf{E}|u_m(s)|_{2\beta}^2 ds.$$

Using (3.120), we obtain for $p = 0, \beta$,

$$\mathbf{E}|v(t)|_p^2 \le Ct^{\frac{\alpha\varepsilon}{2\beta}}h^{4\beta-2p-2\varepsilon}\int_0^t (t-s)^{\frac{\alpha\varepsilon}{2\beta}-1}s^\alpha \frac{1}{\Delta t}\sum_{i=1}^\infty \lambda_i^{\frac{\beta(\alpha+1)}{\alpha}}(\eta_i^m)^2 ds$$
$$\le Ct^{\frac{\alpha\varepsilon}{2\beta}}h^{4\beta-2p-2\varepsilon}\frac{1}{\Delta t}\sum_{i=1}^\infty \lambda_i^{\frac{\beta(\alpha+1)}{\alpha}}(\eta_i^m)^2 B\left(\frac{\alpha\varepsilon}{2\beta}, 1+\alpha\right)$$
$$\le C\varepsilon^{-1}h^{4\beta-2p-2\varepsilon}\frac{1}{\Delta t}\sum_{i=1}^\infty \lambda_i^{\frac{\beta(\alpha+1)}{\alpha}}(\eta_i^m)^2.$$

$$(3.122)$$

The last inequality follows from the fact $B(\frac{\alpha\varepsilon}{2\beta}, 1+\alpha) = \frac{\Gamma(\frac{\alpha\varepsilon}{2\beta})\Gamma(1+\alpha)}{\Gamma(\frac{\alpha\varepsilon}{2\beta}+1+\alpha)}$ and $\Gamma(\frac{\alpha\varepsilon}{2\beta}) \sim \frac{2\beta}{\alpha\varepsilon}$ as $\varepsilon \to 0^+$. Then the desired assertion follows by choosing $\varepsilon = 1/\ell_h$ and the triangle inequality. $\square$

As a simple consequence of Theorem 3.11, Remark 3.2 and Theorem 3.14, we have

Theorem 3.15 *Let u_m be the solution of problem (3.79) with $u_0 \in \dot{U}^q$, $q \in [\beta, 2\beta]$, $v_0 \in \dot{U}^r$, $r \in [0, 2\beta]$, and u_m^h be the solution of (3.97) with $u_0^h = P_h u_0, v_0^h = P_h v_0$. Assume that $\{\sigma_i^m(t)\}$ are uniformly bounded by $|\sigma_i^m(t)| \le \eta_i^m \; \forall t \in [0, T]$. Then with $\ell_h = |\ln h|$,*

$$
\mathbf{E}\|u_m(t) - u_m^h(t)\|^2 + h^{2\beta}\mathbf{E}\|(-\Delta)^{\frac{\beta}{2}}(u_m^h(t) - u_m(t))\|^2
$$
$$
\le C\ell_h^2 h^{4\beta} t^{\frac{\alpha q}{\beta} - 2\alpha}\mathbf{E}|u_0|_q^2 + C\ell_h^2 h^{4\beta} t^{2 - 2\alpha + \frac{\alpha r}{\beta}}\mathbf{E}|v_0|_r^2
$$
$$
+ C\ell_h h^{4\beta}\frac{1}{\Delta t}\sum_{i=1}^{\infty}\lambda_i^{\frac{\beta(\alpha+1)}{\alpha}}(\eta_i^m)^2 \quad \forall t \in [0, T],
$$

provided that the infinite series are convergent, where C is a positive constant independent of Δt and h.

Furthermore, thanks to Theorems 3.9 and 3.15, a space-time error estimate for problem (3.78) follows from the triangle inequality.

Theorem 3.16 *Suppose that the assumptions of Theorem 3.9 hold. Let u be the solution of (3.78) with $u_0 \in \dot{U}^q$, $q \in [\beta, 2\beta]$, $v_0 \in \dot{U}^r$, $r \in [0, 2\beta]$, and u_m^h be the solution of (3.97) with $u_0^h = P_h u_0, v_0^h = P_h v_0$. Then for $\ell_h = |\ln h|$,*

$$
\mathbf{E}\|u(t) - u_m^h(t)\|^2 \le C\sum_{i=1}^{\infty}\lambda_i^{-\frac{2\beta(\alpha-1)}{\alpha}}(\vartheta_i^m)^2
$$
$$
+ C(\Delta t)^2\sum_{i=1}^{\infty}\lambda_i^{-\frac{2\beta(\alpha-1)}{\alpha}}(\zeta_i^m)^2
$$
$$
+ C(\Delta t)^2\sum_{i=1}^{\infty}\lambda_i^{-\frac{\beta(\alpha-2)}{\alpha}}(\eta_i^m)^2 + C\ell_h^2 h^{4\beta} t^{\frac{\alpha q}{\beta} - 2\alpha}\mathbf{E}|u_0|_q^2
$$
$$
+ C\ell_h^2 h^{4\beta} t^{2 - 2\alpha + \frac{\alpha r}{\beta}}\mathbf{E}|v_0|_r^2 + C\ell_h h^{4\beta}\frac{1}{\Delta t}\sum_{i=1}^{\infty}\lambda_i^{\frac{\beta(\alpha+1)}{\alpha}}(\eta_i^m)^2,
$$

provided that the infinite series are all convergent, where C is a positive constant independent of Δt and h.

3.3.2.4   Numerical results

Here we present some numerical tests to verify the theoretical error estimates for the Galerkin FEM. For definiteness, we simulate (3.78), (3.79) and (3.97) with

$$\sigma_i(t) = \frac{1}{i^3},$$

$$\sigma_i^m(t) = \begin{cases} \sigma_i(t), & i \le m, \\ 0, & i > m, \end{cases} \qquad u_0(x) = -4x^2 + 4x, \quad v_0(x) = x.$$

The upper bounds ϑ_i^m, η_i^m and ζ_i^m given in Theorem 3.9 can be chosen as

$$\vartheta_i^m(t) = \begin{cases} 0, & i \le m, \\ \frac{1}{i^3}, & i > m, \end{cases} \qquad \eta_i^m = \zeta_i^m = \frac{1}{i^3}.$$

To obtain the numerical errors and convergence orders, according to the definition of Itô integral, we introduce the reference ("exact") solution u_{ref} defined by

$$
\begin{aligned}
u_{ref} = & \int_0^1 \mathcal{T}_{\alpha,\beta}(t,x,y)u_0(y)dy + \int_0^1 \mathcal{R}_{\alpha,\beta}(t,x,y)v_0(y)dy \\
& + \sum_{m=1}^{M'} \int_0^1 \mathcal{S}_{\alpha,\beta}(t - t_m, x, y) \sum_{i=1}^{N'} \sigma_i(t_m)(\xi_i(t_{m+1}) \\
& - \xi_i(t_m))e_i(y)dy,
\end{aligned}
\tag{3.123}
$$

where $M' = 1000$ with the uniform discretization of the time interval; $N' = 1000$, $\lambda_i = k^2\pi^2$, and $\phi_i(y) = \sqrt{2}\sin k\pi y$.

In our numerical experiments, we first begin with a study of L^2-norm error in the mean-squared sense between the solutions u_{ref} and u_m, which is based on the space and time discretization of the noise. We only need to simulate the stochastic integrals in (3.80) and (3.91). By Theorem 3.9, we need to verify

$$(\mathbf{E}\|u_{ref}(t_{I+1}) - u_m(t_{I+1})\|^2)^{\frac{1}{2}} \le \begin{cases} C(\Delta t)^{\max\{\frac{1}{2}, \alpha - \frac{1}{2} - \varepsilon\}}, & 1 < \alpha \le \frac{3}{2}, \\ C\Delta t, & \frac{3}{2} < \alpha < 2. \end{cases}$$

$$\tag{3.124}$$

Since

$$(\mathbf{E}\|u_{ref}(t_{I+1}) - u_m(t_{I+1})\|^2)^{\frac{1}{2}}$$

$$\approx \left(\frac{1}{M}\sum_{l=1}^{M}\|u_{ref}(t_{I+1},\omega_l) - u_m(t_{I+1},\omega_l)\|^2\right)^{\frac{1}{2}},$$

we take $M = 1000$ as the number of simulation trajectories. The numerical results with $\beta = 0.75$ and $t_{I+1} = T = 1$ are presented in Table 3.9. From Table 3.9, it can be observed that the mean-squared L^2-norm errors converge quickly as Δt decreases and its convergence rates are closely related to the values of α, which agrees well with (3.124). These results confirm the error estimate in Theorem 3.9.

Next, we measure the mean-squared L^2-norm error between u_m and the finite element solution u_m^h. We first divide the unit interval $(0, 1)$ into $N + 1$ equally spaced subintervals with $h = 1/(N + 1)$. Let V_h be the finite element space consisting of continuous piecewise linear polynomials. Notice that the eigenpairs $\{\lambda_j^{h,\beta}, e_j^h(x)\}$ of the one-dimensional discrete fractional Laplacian $(-\Delta_h)^\beta$, defined by (3.96), satisfy

$$((-\Delta_h)^\beta e_j^h, \chi) = \lambda_j^{h,\beta}(e_j^h, \chi) \qquad \forall \chi \in V_h. \qquad (3.125)$$

Then, for $j = 1, 2, \ldots, N$,

$$\lambda_j^{h,\beta} = \sum_{i=1}^{\infty}\lambda_i^\beta(e_j^h, \phi_i)^2. \qquad (3.126)$$

The expressions (3.125) and (3.126) are used in computing the finite element solution of the Galerkin methods through its representation (3.101). To validate the convergence rate 2β of $\|u_m(t_{I+1}) - u_m^h(t_{I+1})\|_{L^2(D,U)}$, we simulate the integral expressions in (3.91) and (3.101) with $\Delta t = 0.01$ and the number of trajectories $M = 500$. The numerical results with $\alpha = 1.5, t_{I+1} = T = 1$ are presented in Table 3.10. From Table 3.10 we can see that, for fixed I, the results of $\|u_m(t_{I+1}) - u_m^h(t_{I+1})\|_{L^2(D,U)}$ converge as h approaches 0, and the corresponding convergence rates have at least an order of $\mathcal{O}(h^{2\beta})$, which in turn justifies the statement of Theorem 3.15.

Combining the data of Tables 3.9 and 3.10 gives the convergence rates, at least $\mathcal{O}((h^{2\beta} + \Delta t)^{\max\{\frac{1}{2}, \alpha - \frac{1}{2} - \varepsilon\}})$ for $1 < \alpha \le 3/2$

Table 3.9 The mean-squared L^2-norm modeling errors and convergence rates of (3.79) with $\beta = 0.75$, $t_{I+1} = 1$ and $M = 1000$

$1/\Delta t$	$\alpha = 1.1$	Rate	$\alpha = 1.25$	Rate	$\alpha = 1.5$	Rate	$\alpha = 1.75$	Rate	$\alpha = 2.0$	Rate
25	1.2567e-2	–	1.5505e-2	–	1.3168e-2	–	9.8503e-3	–	7.1887e-3	–
50	7.6842e-3	0.7097	9.5036e-3	0.7062	7.3208e-3	0.8470	5.1609e-3	0.9325	3.7898e-3	0.9236
100	4.8080e-3	0.6765	5.4252e-3	0.8088	3.7393e-3	0.9693	2.5460e-3	1.0194	1.8082e-3	1.0676
125	4.2249e-3	0.5793	4.6975e-3	0.6454	3.0322e-3	0.9392	2.0506e-3	0.9697	1.4707e-3	0.9258
200	2.9152e-3	0.7895	3.1393e-3	0.8575	2.0008e-3	0.8846	1.3215e-3	0.9348	9.4838e-4	0.9335

Table 3.10 The mean-squared L^2-norm errors and convergence rates of the FEM approximations (3.97) with $\alpha = 1.5$, $t_{I+1} = 1$, $\Delta t = 0.01$ and $M = 500$

$1/h$	$\beta = 0.6$	Rate	$\beta = 0.8$	Rate	$\beta = 1.0$	Rate
10	7.6377e-3	–	6.8670e-3	–	4.0807e-3	–
25	1.3772e-3	1.8695	1.1233e-3	1.9759	5.9384e-4	2.1035
50	3.8494e-4	1.8390	2.7870e-4	2.0109	1.6592e-4	1.8396
75	1.8444e-4	1.8146	1.2644e-4	1.9493	7.3794e-5	1.9983
100	1.0920e-4	1.8221	6.9840e-5	2.0632	4.1522e-5	1.9989

and $\mathcal{O}(\Delta t + h^{2\beta})$ for $3/2 < \alpha < 2$, where the errors, measured by the mean-squared L^2-norm, are the differences between the reference ("exact") solution $u_{ref}(t_{I+1}, \omega_l)$ and the numerical solution $u_m^h(t_{I+1}, \omega_l)$. This means that Theorem 3.16 is valid.

The interesting dynamics of wave propagation governed by (3.79) are shown in Figure 3.3. It is observed with the propagation

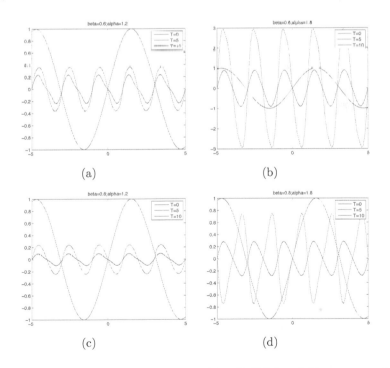

(a) (b)

(c) (d)

Figure 3.3 Wave propagations governed by (3.79) for different α and β. To solve the equation, we take $u_0 = \sin(x)$ and $v_0 = 0$

of the waves that (a) the number of waves per unit distance tends to be fixed; (b) for the same β, the smaller α is, the faster the amplitude decays; and (c) for the same α, the bigger β is, the faster the amplitude attenuates. That is to say, small α and large β implies fast decay of the amplitudes of the waves.

Bibliography

[1] J. Aaronson. *An Introduction to Infinite Ergodic Theory.* American Mathematical Society, Providence, 1997.

[2] M. Abramowitz and I. A. Stegun. *Handbook of Mathematical Functions.* Dover, New York, 1972.

[3] G. Acosta, F. M. Bersetche, and J. P. Borthagaray. A short FE implementation for a 2d homogeneous Dirichlet problem of a fractional Laplacian. *Comput. Math. Appl.*, 74:784–816, 2017.

[4] G. Acosta, F. M. Bersetche, and J. P. Borthagaray. Finite element approximations for fractional evolution problems. *Fract. Calc. Appl. Anal.*, 22:767–794, 2019.

[5] G. Acosta and J. P. Borthagaray. A fractional Laplace equation: Regularity of solutions and finite element approximations. *SIAM J. Numer. Anal.*, 55:472–495, 2017.

[6] R. A. Adams and J. J. F. Fournier. *Sobolev Spaces.* Academic Press, Amsterdam and Boston, 2nd ed., 2003.

[7] E. Aghion, D. A. Kessler, and E. Barkai. From non-normalizable Boltzmann-Gibbs statistics to infinite-ergodic theory. *Phys. Rev. Lett.*, 122:010601, 2019.

[8] P. Allegrini, J. Bellazzini, G. Bramanti, M. Ignaccolo, P. Grigolini, and J. Yang. Scaling breakdown: a signature of aging. *Phys. Rev. E*, 66:015101(R), 2002.

[9] E. Alós, O. Mazet, and D. Nualart. Stochastic calculus with respect to Gaussian processes. *Ann. Probab.*, 29:766–801, 2001.

[10] K. H. Andersen, P. Castiglione, A. Mazzino, and A. Vulpiani. Simple stochastic models showing strong anomalous diffusion. *Eur. Phys. J. B*, 18:447–452, 2000.

[11] R. Anton, D. Cohen, S. Larsson, and X. Wang. Full discretization of semilinear stochastic wave equations driven by multiplicative noise. *SIAM J. Numer. Anal.*, 54:1093–1119, 2016.

[12] D. Applebaum. *Lévy Processes and Stochastic Calculus.* Cambridge University Press, Cambridge, 2009.

[13] D. N. Armstead, B. R. Hunt, and E. Ott. Anomalous diffusion in infinite horizon billiards. *Phys. Rev. E*, 67:021110, 2003.

[14] R. Artuso and G. Cristadoro. Anomalous transport: a deterministic approach. *Phys. Rev. Lett.*, 90:244101, 2003.

[15] R. Artuso, G. Cristadoro, M. D. Esposti, and G. Knight. Sparre-Andersen theorem with spatiotemporal correlations. *Phys. Rev. E*, 89:052111, 2014.

[16] E. Barkai. Fractional Fokker-Planck equation, solution, and application. *Phys. Rev. E*, 63:046118, 2001.

[17] E. Barkai, E. Aghion, and D. A. Kessler. From the area under the Bessel excursion to anomalous diffusion of cold atoms. *Phys. Rev. X*, 4:021036, 2014.

[18] F. Baudoin and M. Hairer. A version of Hörmander's theorem for the fractional Brownian motion. *Probab. Theory Relat. Fields*, 139:373–395, 2007.

[19] A. Baule and R. Friedrich. Joint probability distributions for a class of non-Markovian processes. *Phys. Rev. E*, 71:026101, 2005.

[20] E. Bazhlekova, B. T. Jin, R. Lazarov, and Z. Zhou. An analysis of the Rayleigh-Stokes problem for a generalized second-grade fluid. *Numer. Math.*, 131:1–31, 2015.

[21] P. Bernabo, R. Burioni, S. Lepri, and A. Vezzani. Anomalous transmission and drifts in one-dimensional Lévy structures. *Chaos Soliton. Fract.*, 67:11–19, 2014.

[22] F. Biagini, Y. H. Hu, B. Ø ksendal, and T. Zhang. *Stochastic Calculus for Fractional Brownian Motion and Applications.* Springer, London, 2008.

[23] D. Blömker and A. Jentzen. Galerkin approximations for the stochastic Burgers equation. *SIAM J. Numer. Anal.*, 51:694–715, 2013.

[24] J.-P. Bouchaud and A. Georges. Comment on "Stochastic pathway to anomalous diffusion". *Phys. Rev. A*, 41:1156–1157, 1990.

[25] S. C. Brenner and L. R. Scott. *The Mathematical Theory of Finite Element Methods.* Springer, New York, 3rd edn., 2008.

[26] R. Burioni, L. Caniparoli, and A. Vezzani. Lévy walks and scaling in quenched disordered media. *Phys. Rev. E*, 81:060101, 2010.

[27] S. Burov, J.-H. Jeon, R. Metzler, and E. Barkai. Single particle tracking in systems showing anomalous diffusion: the role of weak ergodicity breaking. *Phys. Chem. Chem. Phys.*, 13.1800 1812, 2011.

[28] A. Cairoli and A. Baule. Anomalous processes with general waiting times: functionals and multipoint structure. *Phys. Rev. Lett.*, 115:110601, 2015.

[29] A. Cairoli and A. Baule. Feynman-Kac equation for anomalous processes with space- and time-dependent forces. *J. Phys. A*, 50:164002, 2017.

[30] S. Carmi and E. Barkai. Fractional Feynman-Kac equation for weak ergodicity breaking. *Phys. Rev. E*, 84:061104, 2011.

[31] S. Carmi, L. Turgeman, and E. Barkai. On distributions of functionals of anomalous diffusion paths. *J. Stat. Phys.*, 141:1071–1092, 2010.

[32] B. A. Carreras, V. E. Lynch, D. E. Newman, and G. M. Zaslavsky. Anomalous diffusion in a running sandpile model. *Phys. Rev. E*, 60:4770–4778, 1999.

[33] P. Castiglione, A. Mazzino, P. Muratore-Ginanneschi, and A. Vulpiani. On strong anomalous diffusion. *Physica D*, 134:75–93, 1999.

[34] A. V. Chechkin, R. Metzler, V. Y. Gonchar, J. Klafter, and L. V. Tanatarov. First passage and arrival time densities for Lévy flights and the failure of the method of images. *J. Phys. A*, 36:L537–L544, 2003.

[35] W. Chen and S. Holm. Fractional Laplacian time-space models for linear and nonlinear lossy media exhibiting arbitrary frequency power-law dependency. *J. Acoust. Soc. Am.*, 115:1424–1430, 2004.

[36] Y. Chen, X. D. Wang, and W. H. Deng. Localization and ballistic diffusion for the tempered fractional Brownian-Langevin motion. *J. Stat. Phys.*, 169:18–37, 2017.

[37] Y. Chen, X. D. Wang, and W. H. Deng. Tempered fractional Langevin-Brownian motion with inverse β-stable subordinator. *J. Phys. A*, 51:495001, 2018.

[38] W. T. Coffey, Y. P. Kalmykov, and J. T. Waldron. *The Langevin Equation*. World Scientific, Singapore, 2004.

[39] D. Cohen and L. Quer-Sardanyons. A fully discrete approximation of the one-dimensional stochastic wave equation. *IMA J. Numer. Anal.*, 36:400–420, 2016.

[40] M. Courbage, M. Edelman, S. M. Saberi Fathi, and G. M. Zaslavsky. Problem of transport in billiards with infinite horizon. *Phys. Rev. E*, 77:036203, 2008.

[41] L. Coutin, D. Nualart, and C. A. Tudor. Tanaka formula for the fractional Brownian motion. *Stoch. Process. Their Appl.*, 94:301–315, 2001.

[42] S. Darses and B. Saussereau. Time reversal for drifted fractional Brownian motion with Hurst index $H > 1/2$. *Electron, J. Probab.*, 12:1181–1211, 2007.

[43] A. Dechant and E. Lutz. Anomalous spatial diffusion and multifractality in optical lattices. *Phys. Rev. Lett.*, 108:230601, 2012.

[44] A. Dechant, E. Lutz, D. A. Kessler, and E. Barkai. Scaling Green-Kubo relation and application to three aging systems. *Phys. Rev. X*, 4:011022, 2014.

[45] L. Decreusefond and A. S. Üstünel. Stochastic analysis of the fractional Brownian motion. *Potential Anal.*, 10:177–214, 1999.

[46] W. H. Deng. Numerical algorithm for the time fractional Fokker-Planck equation. *J. Comput. Phys.*, 227:1510–1522, 2007.

[47] W. H. Deng. Finite element method for the space and time fractional Fokker-Planck equation. *SIAM J. Numer. Anal.*, 47:204–226, 2009.

[48] W. H. Deng and E. Barkai. Ergodic properties of fractional Brownian-Langevin motion. *Phys. Rev. E*, 79:011112, 2009.

[49] W. H. Deng and Z. J. Zhang. *High Accuracy Algorithm for the Differential Equations Governing Anomalous Diffusion*. World Scientific, 2019.

[50] S. I. Denisov, W. Horsthemke, and P. Hänggi. Generalized Fokker-Planck equation: derivation and exact solutions. *Eur. Phys. J. B*, 68:567–575, 2009.

[51] E. Di Nezza, G. Palatucci, and E. Valdinoci. Hitchhiker's guide to the fractional Sobolev spaces. *Bull. Sci. Math.*, 136:521–573, 2012.

[52] Q. Du and T. Y. Zhang. Numerical approximation of some linear stochastic partial differential equations driven by special additive noises. *SIAM J. Numer. Anal.*, 40:1421–1445, 2002.

[53] T. E. Duncan, Y. Z. Hu, and B. Pasik-Duncan. Stochastic calculus for fractional Brownian motion-I. *theory. SIAM J. Control Optim.*, 38:582–612, 2000.

[54] N. T. Dung. Kolmogorov distance between the exponential functionals of fractional Brownian motion. *C. R. Math.*, 357:629–635, 2019.

[55] A. Einstein. Über die von der molekularkinetischen theorie der wärme geforderte bewegung von in ruhenden flüssigkeiten suspendierten teilchen. *Ann. Phys.*, 322:549, 1905.

[56] P. Embrechts and M. Maejima. *Selfsimilar Processes*. Princeton University Press, Princeton, 2002.

[57] P. Flajolet. Singularity analysis and asymptotics of Bernoulli sums. *Theoret. Comput. Sci.*, 215:371–381, 1999.

[58] P. Flandrin. On the spectrum of fractional Brownian motions. *IEEE Trans. Inf. Theory*, 35:197–199, 1992a.

[59] P. Flandrin. Wavelet analysis and synthesis of fractional Brownian motion. *IEEE Trans. Inf. Theory*, 38:910–917, 1992.

[60] H. C. Fogedby. Langevin equations for continuous time Lévy flights. *Phys. Rev. E*, 50:1657–1660, 1994.

[61] R. Friedrich, F. Jenko, A. Baule, and S. Eule. Anomalous diffusion of inertial, weakly damped particles. *Phys. Rev. Lett.*, 96:230601, 2006.

[62] R. Friedrich, F. Jenko, A. Baule, and S. Eule. Exact solution of a generalized Kramers-Fokker-Planck equation retaining retardation effects. *Phys. Rev. E*, 74:041103, 2006.

[63] D. Froemberg and E. Barkai. Time-averaged Einstein relation and fluctuating diffusivities for the Lévy walk. *Phys. Rev. E*, 87:030104(R), 2013.

[64] H. Fujita and T. Suzuki. Evolution problems. In *Finite Element Methods (Part 1)*, volume 2 of *Handbook of Numerical Analysis*, pages 789–928. Elsevier, North Holland, 1991.

[65] N. Gal and D. Weihs. Experimental evidence of strong anomalous diffusion in living cells. *Phys. Rev. E*, 81:020903(R), 2010.

[66] G. H. Gao, Z. Z. Sun, and H. W. Zhang. A new fractional numerical differentiation formula to approximate the Caputo fractional derivative and its applications. *J. Comput. Phys.*, 259:33–50, 2014.

[67] T. Geisel, J. Nierwetberg, and A. Zacherl. Accelerated diffusion in Josephson junctions and related chaotic systems. *Phys. Rev. Lett.*, 54:616–619, 1985.

[68] T. Geisel, A. Zacherl, and G. Radons. Generic $1/f$ noise in chaotic Hamiltonian dynamics. *Phys. Rev. Lett.*, 59:2503–2506, 1987.

[69] H. Geman and M. Yor. Bessel processes, Asian options, and perpetuities. *Math. Finance*, 3:349–375, 1993.

[70] A. Godec and R. Metzler. Finite-time effects and ultraweak ergodicity breaking in superdiffusive dynamics. *Phys. Rev. Lett.*, 110:020603, 2013.

[71] C. Godrèche and J. M. Luck. Statistics of the occupation time of renewal processes. *J. Stat. Phys.*, 104:489–524, 2001.

[72] S. Goldstein. On diffusion by discontinuous movements, and on the telegraph equation. *J. Mech. Appl. Math.*, 6:129–156, 1951.

[73] M. S. Green. Markoff random processes and the statistical mechanics of time dependent phenomena. II. irreversible processes in fluids. *J. Chem. Phys.*, 22:398–413, 1954.

[74] G. Gripenberg and I. Norros. On the prediction of fractional Brownian motion. *J. Appl. Probab.*, 33:400–410, 1996.

[75] G. Grubb. Fractional Laplacians on domains, a development of Hörmander's theory of μ-transmission pseudodifferential operators. *Adv. Math.*, 268:478–528, 2015.

[76] J. W. Haus and K. W. Kehr. Diffusion in regular and disordered lattices. *Phys. Rep.*, 150:263–406, 1987.

[77] Y. He, S. Burov, R. Metzler, and E. Barkai. Random time-scale invariant diffusion and transport coefficients. *Phys. Rev. Lett.*, 101:058101, 2008.

[78] E. Heinsalu, M. Patriarca, I. Goychuk, and P. Hänggi. Use and abuse of a fractional Fokker-Planck dynamics for time-dependent driving. *Phys. Rev. Lett.*, 99:120602, 2007.

[79] E. Heinsalu, M. Patriarca, I. Goychuk, G. Schmid, and P. Hänggi. Fractional Fokker-Planck dynamics: Numerical algorithm and simulations. *Phys. Rev. E*, 73:046133, 2006.

[80] R. Hou and W. H. Deng. Feynman-Kac equations for reaction and diffusion processes. *J. Phys. A*, 51:155001, 2018.

[81] Y. Z. Hu. Probability structure preserving and absolute continuity. *Ann. Inst. Henri Poincare-Probab. Stat.*, 38:557–580, 2002.

[82] Y. Z. Hu and S. G. Peng. Backward stochastic differential equation driven by fractional Brownian motion. *SIAM J. Control Optim.*, 48:1675–1700, 2009.

[83] K. Itô. Stochastic differential equations in a differentiable manifold. *Nagoya Math. J.*, 1:35–47, 1950.

[84] B. T. Jin, R. Lazarov, J. Pasciak, and Z. Zhou. Error analysis of a finite element method for the space-fractional parabolic equation. *SIAM J. Numer. Anal.*, 52:2272–2294, 2014.

[85] B. T. Jin, R. Lazarov, J. Pasciak, and Z. Zhou. Error analysis of semidiscrete finite element methods for inhomogeneous time-fractional diffusion. *IMA J. Numer. Anal.*, 35:561–582, 2015.

[86] B. T. Jin, R. Lazarov, and Z. Zhou. Error estimates for a semidiscrete finite element method for fractional order parabolic equations. *SIAM J. Numer. Anal.*, 51:445–466, 2013.

[87] B. T. Jin, R. Lazarov, and Z. Zhou. An analysis of the L1 scheme for the subdiffusion equation with nonsmooth data. *IMA J. Numer. Anal.*, 36:197–221, 2015.

[88] B. T. Jin, R. Lazarov, and Z. Zhou. Two fully discrete schemes for fractional diffusion and diffusion-wave equations with nonsmooth data. *SIAM J. Sci. Comput.*, 38:A146–A170, 2016.

[89] B. T. Jin, R. Lazarov, and Z. Zhou. Numerical methods for time-fractional evolution equations with nonsmooth data:

a concise overview. *Comput. Methods Appl. Mech. Engrg.*, 346:332–358, 2019.

[90] B. T Jin, B. Y. Li, and Z. Zhou. Correction of high-order BDF convolution quadrature for fractional evolution equations. *SIAM J. Sci. Comput.*, 39:A3129–A3152, 2017.

[91] Y.-J. Jung, E. Barkai, and R. J. Silbey. Lineshape theory and photon counting statistics for blinking quantum dots: a Lévy walk process. *Chem. Phys.*, 284:181–194, 2002.

[92] M. Kac. On distributions of certain Wiener functionals. *Trans. Amer. Math. Soc.*, 65:1–13, 1949.

[93] M. Kamrani and S. M. Hosseini. The role of coefficients of a general SPDE on the stability and convergence of a finite difference method. *J. Comput. Appl. Math.*, 234:1426–1434, 2010.

[94] D. A. Kessler and E. Barkai. Infinite covariant density for diffusion in logarithmic potentials and optical lattices. *Phys. Rev. Lett.*, 105:120602, 2010.

[95] D. A. Kessler and E. Barkai. Theory of fractional Lévy kinetics for cold atoms diffusing in optical lattices. *Phys. Rev. Lett.*, 108:230602, 2012.

[96] A. A. Kilbas, H. M. Srivastava, and J. J. Trujillo. *Theory and Applications of Fractional Differential Equations*. Elsevier, Amsterdam, 2006.

[97] J. Klafter, A. Blumen, and M. F. Shlesinger. Stochastic pathway to anomalous diffusion. *Phys. Rev. A*, 35:3081–3085, 1987.

[98] J. Klafter and I. M. Sokolov. *First Steps in Random Walks From Tools to Applications*. Oxford University Press, New York, 2011.

[99] J. Klafter and G. Zumofen. Lévy statistics in a Hamiltonian system. *Phys. Rev. E*, 49:4873–4877, 1994.

[100] N. Korabel and E. Barkai. Anomalous infiltration. *J. Stat. Mech.*, 2011:P05022, 2011.

[101] T. Koren, M. A. Lomholt, A. V. Chechkin, J. Klafter, and R. Metzler. Leapover lengths and first passage time statistics for Lévy flights. *Phys. Rev. Lett.*, 99:160602, 2007.

[102] S. C. Kou and X. S. Xie. Generalized Langevin equation with fractional Gaussian noise: subdiffusion within a single protein molecule. *Phys. Rev. Lett.*, 93:180603, 2004.

[103] M. Kovács and J. Printems. Strong order of convergence of a fully discrete approximation of a linear stochastic Volterra type evolution equation. *Math. Comput.*, 83:2325–2346, 2014.

[104] R. Kubo. Statistical-mechanical theory of irreversible processes. I. general theory and simple applications to magnetic and conduction problems. *J. Phys. Soc. Jpn.*, 12:570–586, 1957.

[105] R. Kubo. The fluctuation-dissipation theorem. *Rep. Prog. Phys.*, 29:255–284, 1966.

[106] J. Lamperti. An occupation time theorem for a class of stochastic processes. *Trans. Am. Math. Soc.*, 88:380–387, 1958.

[107] P. Langevin. On the theory of Brownian motion. *C. R. Acad. Sci.*, 146:530–533, 1908.

[108] T. A. M. Langlands and B. I. Henry. The accuracy and stability of an implicit solution method for the fractional diffusion equation. *J. Comput. Phys.*, 205:719–736, 2005.

[109] A. Laptev. Dirichlet and Neumann eigenvalue problems on domains in Euclidean spaces. *J. Funct. Anal.*, 151:531–545, 1997.

[110] N. Leibovich and E. Barkai. Infinite ergodic theory for heterogeneous diffusion processes. *Phys. Rev. E*, 99:042138, 2019.

[111] P. Lévy. Sur certains processus stochastiques homogènes. *Compos. Math.*, 7:283–339, 1939.

[112] L. Lewin. *Polylogarithms and Associated Functions*. North-Holland, New York and Oxford, 1981.

[113] C. P. Li and H. F. Ding. Higher order finite difference method for the reaction and anomalous-diffusion equation. *Appl. Math. Model.*, 38:3802–3821, 2014.

[114] P. Li and S. T. Yau. On the Schrödinger equation and the eigenvalue problem. *Comm. Math. Phys.*, 88:309–318, 1983.

[115] Y. J. Li, Y. J. Wang, and W. H. Deng. Galerkin finite element approximations for stochastic space-time fractional wave equations. *SIAM J. Numer. Anal.*, 55:3173–3202, 2017.

[116] Y. M. Lin and C. J. Xu. Finite difference/spectral approximations for the time-fractional diffusion equation. *J. Comput. Phys.*, 225:1533–1552, 2007.

[117] J. L. Lions and E. Magenes. *Non-Homogeneous Boundary Value Problems and Applications.* Springer Berlin Heidelberg, Berlin, Heidelberg, 1972.

[118] F. Liu, V. Anh, and I. Turner. Numerical solution of the space fractional Fokker–Planck equation. *J. Comput. Appl. Math.*, 166:209–219, 2004.

[119] X. Liu and W. H. Deng. Numerical approximation for fractional diffusion equation forced by a tempered fractional Gaussian noise. *J. Sci. Comput.*, 84:1–28, 2020.

[120] X. Liu and W. H. Deng. Higher order approximation for stochastic space fractional wave equation forced by an additive space-time Gaussian noise. *J. Sci. Comput.*, 87:1–29, 2021.

[121] Z. H. Liu and Z. H. Qiao. Strong approximation of monotone stochastic partial differential equations driven by white noise. *IMA J. Numer. Anal.*, 40:1074–1093, 2020.

[122] C. Lubich. Convolution quadrature and discretized operational calculus. I. *Numer. Math.*, 52:129–145, 1988.

[123] C. Lubich. Convolution quadrature and discretized operational calculus. II. *Numer. Math.*, 52:413–425, 1988.

[124] C. Lubich, I. H. Sloan, and V. Thomée. Nonsmooth data error estimates for approximations of an evolution equation

with a positive-type memory term. *Math. Comp.*, 65:1–18, 1996.

[125] E. Lutz. Fractional Langevin equation. *Phys. Rev. E*, 64:051106, 2001.

[126] M. Magdziarz, A. Weron, and J. Klafter. Equivalence of the fractional Fokker-Planck and subordinated Langevin equations: the case of a time-dependent force. *Phys. Rev. Lett.*, 101:210601, 2008.

[127] S. N. Majumdar. Brownian functionals in physics and computer science. *Curr. Sci.*, 89:2076–2092, 2005.

[128] S. Marksteiner, K. Ellinger, and P. Zoller. Anomalous diffusion and Lévy walks in optical lattices. *Phys. Rev. A*, 53:3409–3430, 1996.

[129] A. M. Mathai and R. K. Saxena. *The H-Function with Applications in Statistics and Other Disciplines*. Wiley Eastern, New Delhi, 1978.

[130] A. M. Mathai, R. K. Saxena, and H. J. Haubold. *The H-Function, Theory and Applications*. Springer, Berlin, 2009.

[131] M. M. Meerschaert and F. Sabzikar. Tempered fractional Brownian motion. *Stat. Probab. Lett.*, 83:2269–2275, 2013.

[132] M. M. Meerschaert and F. Sabzikar. Stochastic integration for tempered fractional Brownian motion. *Stoch. Process. Appl.*, 124:2363–2387, 2014.

[133] M. M. Meerschaert, H. P. Scheffler, and C. Tadjeran. Finite difference methods for two-dimensional fractional dispersion equation. *J. Comput. Phys.*, 211:249–261, 2006.

[134] M. M. Meerschaert, R. L. Schilling, and A. Sikorskii. Stochastic solutions for fractional wave equations. *Nonlinear Dyn.*, 80:1685–1695, 2015.

[135] R. Metzler, J.-H. Jeon, A. G. Cherstvy, and E. Barkai. Anomalous diffusion models and their properties: non-stationarity, non-ergodicity, and ageing at the centenary of single particle tracking. *Phys. Chem. Chem. Phys.*, 16:24128–24164, 2014.

[136] R. Metzler and J. Klafter. The random walk's guide to anomalous diffusion: a fractional dynamics approach. *Phys. Rep.*, 339:1–77, 2000.

[137] P. Meyer, E. Barkai, and H. Kantz. Scale-invariant Green-Kubo relation for time-averaged diffusivity. *Phys. Rev. E*, 96:062122, 2017.

[138] Y. S. Mishura. *Stochastic Calculus for Fractional Brownian Motion and Related Processes*. Springer, Berlin, 2008.

[139] T. Miyaguchi, T. Akimoto, and E. Yamamoto. Langevin equation with fluctuating diffusivity: a two-state model. *Phys. Rev. E*, 94:012109, 2016.

[140] D. Molina-Garcia, T. Sandev, H. Safdari, G. Pagnini, A. Chechkin, and R. Metzler. Crossover from anomalous to normal diffusion: truncated power-law noise correlations and applications to dynamics in lipid bilayers. *New J. Phys.*, 20:103027, 2018.

[141] E. W. Montroll and G. H. Weiss. Random walks on lattices. II. *J. Math. Phys.*, 6:167–181, 1965.

[142] D. E. Newman, R. Sánchez, B. A. Carreras, and W. Ferenbaugh. Transition in the dynamics of a diffusive running sandpile. *Phys. Rev. Lett.*, 88:204304, 2002.

[143] I. Norros, E. Valkeila, and J. Virtamo. An elementary approach to a Girsanov formula and other analytic results on fractional Brownian motions. *Bernoulli*, 5:571–587, 1999.

[144] B. Øksendal. *Stochastic Differential Equations*. Springer-Verlag, Berlin, 2005.

[145] V. V. Palyulin, G. Blackburn, M. A. Lomholt, N. W. Watkins, R. Metzler, R. Klages, and A. V Chechkin. First passage and first hitting times of Lévy flights and Lévy walks. *New J. Phys.*, 21:103028, 2019.

[146] V. V. Palyulin, A. V. Chechkin, R. Klages, and R. Metzler. Search reliability and search efficiency of combined Lévy-Brownian motion: long relocations mingled with thorough local exploration. *J. Phys. A*, 49:394002, 2016.

[147] V. V. Palyulin, A. V. Chechkin, and R. Metzler. Space-fractional Fokker-Planck equation and optimization of random search processes in the presence of an external bias. *J. Stat. Mech*, 2014:P11031, 2014.

[148] A. S. Pikovsky. Statistical properties of dynamically generated anomalous diffusion. *Phys. Rev. A*, 43:3146–3148, 1991.

[149] A. Piryatinska, A. I. Saichev, and W. A. Woyczynski. Models of anomalous diffusion: the subdiffusive case. *Physica A*, 349:375–420, 2005.

[150] I. Podlubny. *Fractional Differential Equations*. Academic Press, San Diego, 1999.

[151] G. D. Prato and J. Zabczyk. *Stochastic Equations in Infinite Dimensions*. Cambridge University Press, Cambridge, 2nd edn., 2014.

[152] A. Rebenshtok, S. Denisov, P. Hänggi, and E. Barkai. Infinite densities for Lévy walks. *Phys. Rev. E*, 90:062135, 2014.

[153] A. Rebenshtok, S. Denisov, P. Hänggi, and E. Barkai. Non-normalizable densities in strong anomalous diffusion: beyond the central limit theorem. *Phys. Rev. Lett.*, 112:110601, 2014.

[154] A. Rebenshtok, S. Denisov, P. Hänggi, and E. Barkai. Complementary densities of Lévy walks: typical and rare fluctuations. *Math. Model. Nat. Phenom.*, 11:76–106, 2016.

[155] S. Redner. *A Guide to First-Passage Processes*. Cambridge University Press, Cambridge, 2001.

[156] H. Risken. *The Fokker-Planck Equation*. Springer-Verlag, Berlin, 1989.

[157] F. Sabzikar and D. Surgailis. Tempered fractional Brownian and stable motions of second kind. *Statist. Probab. Lett.*, 132:17–27, 2018.

[158] D. P. Sanders and H. Larralde. Occurrence of normal and anomalous diffusion in polygonal billiard channels. *Phys. Rev. E*, 73:026205, 2006.

[159] M. J. Saxton and K. Jacobson. Single-particle tracking: applications to membrane dynamics. *Annu. Rev. Biophys. Biomol. Struct.*, 26:373–399, 1997.

[160] M. Schmiedeberg, V. Yu Zaburdaev, and H. Stark. On moments and scaling regimes in anomalous random walks. *J. Stat. Mech.*, 2009: P12020, 2009.

[161] J. H. P. Schulz and E. Barkai. Fluctuations around equilibrium laws in ergodic continuous-time random walks. *Phys. Rev. E*, 91:062129, 2015.

[162] G. J. Shen, L. W. Xia, and D. J. Zhu. A strong convergence to the tempered fractional Brownian motion. *Comm. Statist. Theory Methods*, 46:4103–4118, 2017.

[163] S. Shklyar, G. Shevchenko, Y. Mishura, V. Doroshenko, and O. Banna. Approximation of fractional Brownian motion by martingales. *Methodol. Comput. Appl. Probab.*, 16:539–560, 2014.

[164] M. F. Shlesinger, G. M. Zaslavsky, and U. Frisch. *Lévy Flights and Related Topics*. Springer-Verlag, Berlin, 1995.

[165] I. M. Sokolov, A. Blumen, and J. Klafter. Linear response in complex systems: CTRW and the fractional Fokker-Planck equations. *Physica A*, 302:268–278, 2001.

[166] T. H. Solomon, E. R. Weeks, and H. L. Swinney. Observation of anomalous diffusion and Lévy flights in a two-dimensional rotating flow. *Phys. Rev. Lett.*, 71:3975–3978, 1993.

[167] M. S. Song, H. C. Moon, J.-H. Jeon, and H. Y. Park. Neuronal messenger ribonucleoprotein transport follows an aging Lévy walk. *Nat. Commun.*, 9:344–351, 2018.

[168] H. M. Srivastava, K. C. Gupta, and S. P. Goyal. *The H-Functions of One and Two Variables with Applications*. South Asian Publishers, New Delhi, 1982.

[169] H. M. Srivastava and B. R. K. Kashyap. *Special Functions in Queuing Theory and Related Stochastic Processes*. Academic Press, New York, 1982.

[170] W. A. Strauss. *Partial Differential Equations: An Introduction*. Wiley, New York, 2008.

[171] T. L. Szabo. Time domain wave equations for lossy media obeying a frequency power law. *J. Acoust. Soc. Am.*, 96:491–500, 1994.

[172] G. I. Taylor. Diffusion by continuous movements. *Proc. London Math. Soc.*, s2-20:196–212, 1922.

[173] M. Thaler and R. Zweimüller. Distributional limit theorems in infinite ergodic theory. *Probab. Theory Rel.*, 135:15–52, 2006.

[174] V. Thomée. *Galerkin Finite Element Methods for Parabolic Problems*. Springer-Verlag, Berlin, 2nd ed., 2006.

[175] L. Turgeman, S. Carmi, and E. Barkai. Fractional Feynman-Kac equation for non-Brownian functionals. *Phys. Rev. Lett.*, 103:190201, 2009.

[176] M. Vahabi, J. H. P. Schulz, B. Shokri, and R. Metzler. Area coverage of radial Lévy flights with periodic boundary conditions. *Phys. Rev. E*, 87:042136, 2013.

[177] M. von Smoluchowski. Zur kinetischen theorie der brownschen molekularbewegung und der suspensionen. *Ann. Phys.*, 326:756–780, 1906.

[178] G. Wang, M. Zeng, and B. Guo. Stochastic Burgers' equation driven by fractional Brownian motion. *J. Math. Anal. Appl.*, 371:210–222, 2010.

[179] W. L. Wang and W. H. Deng. Aging Feynman-Kac equation. *J. Phys. A*, 51:015001, 2018.

[180] W. L. Wang, J. H. P. Schulz, W. H. Deng, and E. Barkai. Renewal theory with fat-tailed distributed sojourn times: typical versus rare. *Phys. Rev. E*, 98:042139, 2018.

[181] X. Wang, S. Gan, and J. Tang. Higher order strong approximations of semilinear stochastic wave equation with additive space-time white noise. *SIAM J. Sci. Comput.*, 36:A2611–A2632, 2014.

[182] X. D. Wang, Y. Chen, and W. H. Deng. Feynman-Kac equation revisited. *Phys. Rev. E*, 98:052114, 2018.

[183] X. D. Wang, Y. Chen, and W. H. Deng. Aging two-state process with Lévy walk and Brownian motion. *Phys. Rev. E*, 100:012136, 2019.

[184] X. D. Wang, Y. Chen, and W. H. Deng. Strong anomalous diffusion in two-state process with Lévy walk and Brownian motion. *Phys. Rev. Res.*, 2:013102, 2020.

[185] X. D. Wang, W. H. Deng, and Y. Chen. Ergodic properties of heterogeneous diffusion processes in a potential well. *J. Chem. Phys.*, 150:164121, 2019.

[186] D. Wirtz. Particle-tracking microrheology of living cells: principles and applications. *Annu. Rev. Biophys.*, 38:301–326, 2009.

[187] K. B. Wolf. *Integral Transforms in Science and Engineering*. Plenum Press, New York, 1979.

[188] X. C. Wu, W. H. Deng, and E. Barkai. Tempered fractional Feynman-Kac equation: theory and examples. *Phys. Rev. E*, 93:032151, 2016.

[189] P. B. Xu and W. H. Deng. Fractional compound Poisson processes with multiple internal states. *Math. Model. Nat. Phenom*, 13:10, 2018.

[190] P. B. Xu and W. H. Deng. Lévy walk with multiple internal states. *J Stat Phys*, 173:1598–1613, 2018.

[191] A. C. Yadav, R. Ramaswamy, and D. Dhar. Power spectrum of mass and activity fluctuations in a sandpile. *Phys. Rev. E*, 85:061114, 2012.

[192] Y. B. Yan. Galerkin finite element methods for stochastic parabolic partial differential equations. *SIAM J. Numer. Anal.*, 43:1363–1384, 2005.

[193] L. Yang and Y. Z. Zhang. Convergence of the spectral Galerkin method for the stochastic reaction-diffusionadvection equation. *Math. Anal. Appl.*, 446:1230–1254, 2017.

[194] A. Yao, M. Tassieri, M. Padgett, and J. Cooper. Microrheology with optical tweezers. *Lab Chip*, 9:2568–2575, 2009.

[195] H. Yoo. Semi-discretization of stochastic partial differential equations on $\mathbb{R}^1$ by a finite-difference method. *Math. Comput.*, 69:653–666, 2000.

[196] M. Yor. *Exponential Functionals of Brownian Motion and Related Processes.* Springer, Berlin, 2000.

[197] V. Zaburdaev, S. Denisov, and J. Klafter. Lévy walks. *Rev. Mod. Phys.*, 87:483–530, 2015.

[198] G. Zumofen and J. Klafter. Scale-invariant motion in intermittent chaotic systems. *Phys. Rev. E*, 47:851–863, 1993.

Index

Printed and bound by CPI Group (UK) Ltd, Croydon, CR0 4YY

17/10/2024

01775680-0010